MOLECULAR
BIOLOGY
INTELLIGENCE
UNIT

MOLECULAR BIOLOGY OF DESMOSOMES AND HEMIDESMOSOMES

Jane E. Collins
University of Southampton
Southampton, United Kingdom

David R. Garrod
University of Manchester
Manchester, United Kingdom

R.G. LANDES COMPANY
AUSTIN

Molecular Biology Intelligence Unit
MOLECULAR BIOLOGY OF DESMOSOMES AND HEMIDESMOSOMES
R.G. LANDES COMPANY
Austin

CRC Press is the exclusive worldwide distributor of publications
of the Molecular Biology Intelligence Unit.
CRC Press, 2000 Corporate Blvd., NW, Boca Raton, FL 33431. Phone: 407/994-0555.

Submitted: September 1994
Published: November 1994

Please address all inquiries to the Publisher:
R.G. Landes Company, 909 Pine Street, Georgetown, TX 78626
or
P.O. Box 4858, Austin, TX 78765

Phone: 512/ 863 7762; FAX: 512/ 863 0081

ISBN 1-57059-197-0
CATALOG # LN9197

Library of Congress Cataloging-in-Publication Data

Molecular biology of desmosomes and hemidesmosomes / [edited by] Jane
E. Collins, David R. Garrod.
p. cm. -- (Molecular biology intelligence unit)
Includes bibliographical references and index.
ISBN 1-57059-197-0
1. Desmosomes. 2. Hemidesmosomes. I. Collins, Jane E.
II. Garrod, David R. III. Series.
QH603.C4M65 1994
611'.0187--dc20 94-38446
CIP

CONTENTS

EDITORS

Jane E. Collins
Department of Biology
School of Biological Sciences
University of Southampton
Southampton, UK
Chapter 2

David R. Garrod
Cancer Research Campaign Epithelial Morphogenesis Research Group
School of Biological Sciences
University of Manchester
Manchester, UK
Chapter 2

CONTRIBUTORS

John D. Aplin
Research Floor
St. Mary's Hospital
Manchester, UK
Chapter 6

Joachim Arnemann
Institut für Humangenetick
Universitätsklinik
Frankfurt, Germany
Chapter 1

Scott E. Baker
Cell and Molecular Biology
Northwestern University Medical School
Chicago, Illinois, USA
Chapter 5

Elayne A. Bornslaeger
Northwestern University Medical School
Departments of Pathology, Dermatology and the R H Lurie Cancer Center
Chicago, Illinois, USA
Chapter 3

Roger S. Buxton
Laboratory of Eukaryotic Molecular Genetics
National Institute for Medical Research
London, UK
Chapter 1

Heather J. Church
Department of Obstetrics and Gynaecology and School of Biological Sciences
University of Manchester
Manchester, UK
Chapter 6

Pamela Cowin
Department of Cell Biology
Ronald O Perelman Department of Dermatology and Kaplan Cancer Center
New York University Medical Center
New York, New York, USA
Chapter 4

Kathleen J. Green
Northwestern University Medical School
Departments of Pathology, Dermatology and the R H Lurie Cancer Center
Chicago, Illinois, USA
Chapter 3

Susan B. Hopkinson
Cell and Molecular Biology
Northwestern University Medical School
Chicago, Illinois, USA
Chapter 5

Jonathan C.R. Jones
Cell and Molecular Biology
Northwestern University Medical School
Chicago, Illinois, USA
Chapter 5

Sarolta Kárpáti
Dermatology Branch
National Institutes of Health
Bethesda, Maryland, USA
Chapter 7

Ian A. King
Laboratory of Eukaryotic Molecular Genetics
National Institute for Medical Research
London, UK
Chapter 1

Andrew P. Kowalczyk
Northwestern University Medical School
Departments of Pathology, Dermatology and the R H Lurie Cancer Center
Chicago, Illinois, USA
Chapter 3

Mikayo Langhofer
Cell and Molecular Biology
Northwestern University Medical School
Chicago, Illinois, USA
Chapter 5

Anthony I. Magee
Laboratory of Eukaryotic Molecular Genetics
National Institute for Medical Research
London, UK
Chapter 1

Helena L. Palka
Northwestern University Medical School
Departments of Pathology, Dermatology and the R H Lurie Cancer Center
Chicago, Illinois, USA
Chapter 3

Thaddeus S. Stappenbeck
Northwestern University Medical School
Departments of Pathology, Dermatology and the R H Lurie Cancer Center
Chicago, Illinois, USA
Chapter 3

John R. Stanley
Dermatology Branch
National Institutes of Health
Bethesda, Maryland, USA
Chapter 7

PREFACE

Desmosomes and hemidesmosomes are cell junctions that mediate adhesion mainly in epithelia. Desmosomes join cells to each other whereas hemidesmosomes are cell-matrix junctions concerned with anchoring cells to their underlying matrix. These junctions are macromolecular protein complexes which are considered to provide vital links in the structural framework of tissues by their interactions with the intermediate filament cytoskeleton.

Desmosomes and hemidesmosomes appear ultrastructurally to have a common feature, namely the presence of dense membrane-associated plaques. Desmosomes are comprised of two plaques each contributed by an apposing cell, whereas hemidesmosomes, as the name suggests, have one plaque which is adjacent to the extracellular matrix. Thus to researchers who observed these structures it was logical, a hemidesmosome was half a desmosome!

This ultrastructural similarity was misleading. Molecular analysis using antibodies and cDNA cloning and sequencing has revealed that the protein constituents of these structures are quite different. The adhesion molecules of the desmosome belong to the cadherin superfamily of cell adhesion molecules whereas adhesion at the hemidesmosome involves members of the integrin superfamily. The plaque proteins are also different. In addition, it is interesting that there may be analogous proteins in the junctions performing similar roles which have regions of structural conservation relating to function, for example desmoplakin, a cytoplasmic protein found in desmosomes, has homology to bullous pemphigoid antigen-1, a plaque protein in hemidesmosomes. These similarities may relate to the common linkage to intermediate filaments and associated proteins. So the differences and the similarities may provide clues to important functional mechanisms used by these proteins.

This book provides a unique collection of articles from groups active in studying the molecular regulation of desmosomes and hemidesmosomes. Molecular genetics provides the basis for elegant approaches to analyzing the many interactions and control mechanisms involved in the assembly and function of proteins. This is particularly relevant to these structures which have proved resistant to mild solubilization techniques. The important questions relate to function. In tissues, cells need to be integrated into cohesive units to provide structural integrity. When these adhesion systems fail the consequences are disastrous as seen in blistering skin diseases such as pemphigus vulgaris where there is a loss of adhesion which appears to involve desmosomes, or in bullous pemphigoid where a lesion occurs at the level of hemidesmosomes. Therefore adhesion mechanisms need to be strong yet flexible to allow cell movement, for example during wound healing or when skin cells move out of the basal layer of the epidermis. These observations prompt questions about how adhesion is regulated. In addition, adhesion mechanisms may transduce information to the cell interior. What does adhesion tell cells about their surroundings and does this influence subsequent decisions about proliferation, differentiation or behavior? The research groups who have contributed to this volume are currently engaged in trying to find answers to these and many other questions, all of which promise to provide very exciting developments in the future.

Jane E. Collins and David R. Garrod

CHAPTER 1

DESMOSOMAL GENES

Roger S. Buxton, Anthony I. Magee, Ian A. King, Joachim Arnemann

Many of the genes coding for proteins of cell junctions appear to have been duplicated and have subsequently evolved so that they are specific for a particular type of junction; thus analogous but distinct proteins are present in for example, the desmosome and the hemidesmosome or in adherens junctions, and in basal desmosomes compared with suprabasal desmosomes. The desmosomal genes can therefore be grouped into a number of gene families, such as the cadherin family, the desmoplakin family and the plakoglobin family. Presumably the structure and expression of the analogous proteins within these families differ in subtle and necessary ways to fulfill their function in the particular location where they are situated. The aim of much current research is directed towards explaining what these functions are and how these proteins fulfill them. This chapter will provide a summary of the present state of knowledge of the human, bovine, murine and canine desmosomal genes.

The cloning of the cDNAs coding for the proteins of the desmosome has been crucial in providing a complete picture of the molecular components of this class of cell junction and has revealed a more subtle and complicated structure than was evident solely from experiments using antibodies as probes. In particular it has demonstrated that the putative adhesive proteins of the desmosome, the desmocollins and desmogleins, are present not as single protein species but as a number of distinct isoforms, a conclusion which would have been very difficult to reach if protein purification experiments alone had been performed.

The ability to isolate the desmosomal proteins in a reasonably pure form and their use to raise specific antibodies has, however, been central to the cDNA cloning. Moreover the evolutionary conservation of structure has meant that antibodies raised against bovine proteins, which are the most accessible source of desmosomes, have mostly reacted against the proteins of human and other species and this has aided the process of cDNA cloning. As a result of these experiments, therefore, human as well as bovine cDNAs have been isolated for all of the major proteins of

Molecular Biology of Desmosomes and Hemidesmosomes, edited by Jane E. Collins and David R. Garrod.

the desmosome, and these have been used to map the location of the genes to human and in some cases mouse chromosomes. Map locations are a necessary prerequisite for judging whether these genes are candidates for any diseases, either human or murine; at present no inherited diseases have been associated with mutations in the desmosomal genes although some of the proteins have been shown to be the autoantigens in certain autoimmune diseases of the skin (see ref. 1 and chapter 7 for reviews).

CLONING OF cDNAs: PROTEIN STRUCTURE

The Desmocollins and Desmogleins: Cadherins of the Desmosome

The desmocollins have been known from immunoblotting and immunoprecipitation experiments as a pair of proteins of approximately 94 kD and 110 kD which are related immunologically and by peptide analysis. They were long suspected of being related to the cadherins on account of being similarly sized transmembrane glycoproteins which were proteolytically processed[2] and more recent work confirmed the transmembrane organization and described amino acid terminal sequence homology with the classic cadherins (E-, P- and N-cadherin).[3] Their involvement in cell-cell adhesion was inferred from the inhibition of desmosomal plaque formation in MDBK cells by Fab' fragments of polyclonal antibodies.[4] Initial reports of the cDNA cloning of the desmocollins, by immunoscreening of phage λ expression libraries constructed with mRNA from cultured human keratinocytes or from bovine snout, demonstrated that these proteins were indeed members of the cadherin superfamily of cell adhesion molecules.[5-8] Thus desmocollins, from both human and bovine sources, like the classical cadherins, have four external repeated domains containing potential Ca^{2+}-binding sites, an extracellular anchor domain, a transmembrane domain and a cytoplasmic domain slightly larger (M_r = 19,885 in human DSC2) than that in, for example, E-cadherin (M_r = 16,971), and a similar size for the mature protein (84,633 for human DSC2 and 80,228 for mouse E-cadherin) (see Fig. 1.1). The explanation of the observed dual desmocollin protein bands lies in an alternative splicing of the initial mRNA transcript.[5,8] The 'b' form differs from the 'a' form in having an extra 46 bp exon containing an in-frame stop codon, resulting in a smaller cytoplasmic domain (M_r = 13,699 as against 19,885 for human DSC2), a different C-terminus of 11 amino acids and loss of a potential serine phosphorylation site, in line with the knowledge that the 'a' form but not the 'b' form is phosphorylated in its cytoplasmic domain in MDCK cells.[3] The small difference between the two message sizes is not distinguishable on most agarose gels and is thus compatible with the single band seen on a Northern blot. No reason has yet been found for the existence of the two alternatively spliced desmocollins although it could be related to their differential phosphorylation. Experiments with chimeric molecules having a connexin extracellular domain and a desmocollin cytoplasmic domain demonstrated that only chimeras with the 'a' form cytoplasmic domain supported plaque assembly when transfected into A431 cells.[9]

These desmosomal cadherins represent a family within the cadherin superfamily; thus human DSC2 has 35% identity with human N-cadherin, which is less than the 45-50% observed when comparing classic cadherins to one another. The cytoplasmic domain of human DSC2 has only about 30% identity with the cadherin cytoplasmic domains, whereas the classic cadherins exhibit 90% identity in their cytoplasmic domains which are essential for interactions with the catenins and the microfilament network. Additionally, one cytoplasmic region of 45 residues in the desmocollins is totally missing from the cadherins; this might be responsible for the selective association with the intermediate filaments or a component of the cytoplasmic plaque.

The other type of desmosomal glycoprotein, the desmogleins, also have homology to the cadherins.[10-14] Desmogleins DSG2 and DSG3 have an extracellular anchor domain similar to that in other cadherins but DSG1 has an extracellular anchor that differs considerably from the desmocollins and from the classical cadherins and could be the site of O-linked glycosylation. All desmogleins also have much larger cytoplasmic domains extending beyond the homologous region.[10,12,14] In

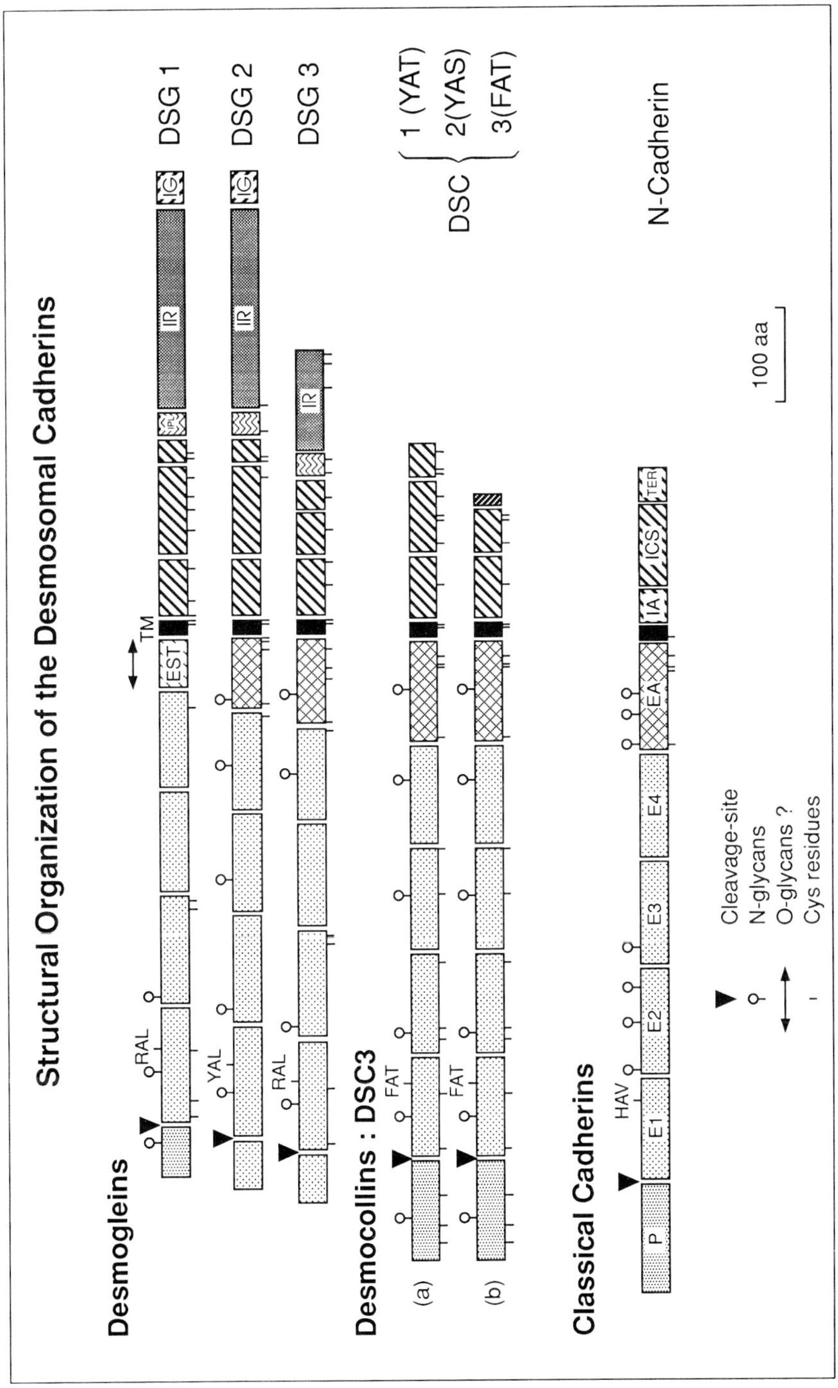

Fig. 1.1. Structural organization of the human desmogleins and desmocollins compared with human N-cadherin. The mature cleavage site is shown by (▼). The sequences in the extracellular domains which have the consensus for N-linked glycosylation are shown by (⚲). The extracellular Ser/Thr-rich domain in DSG1 (EST) is the potential site for O-linked glycosylation(⟷). Cys residues within the sequence are underscored ('). Domains have been aligned according to reference 8. P represents the pro-sequence of each protein; E1, E2, E3 and E4 are the four blocks with repeating units in the extracellular part of the protein; EA, the external anchor domain; TM, the transmembrane region; IA, the intracellular anchor domain; ICS, the intracellular cadherin-type segment; TER, the terminal cadherin homology; IPL, the proline-rich linker; IR, the intracellular repeat domain of the desmogleins; IG, the intracellular Gly and Ser rich region. The alternatively spliced ('a' and 'b' forms) of the desmocollins are shown, resulting in the unique C-terminus of the 'b' form (C_b). It should be noted that the sites for N-linked glycosylation are the same for DSC2 and DSC3 but differ in DSC1.

DSG1 this 282 residue extension contains five repeats of a 29 ± 4 residue sequence with a core consensus of N-V-V/I-V-T-E-R/S-V-I/V, which is predicted to have an anti-parallel β-sheet structure, followed by a glycine-rich sequence. The repeat sequence is diagnostic of desmogleins and occurs six times in DSG2, five times in DSG1 and only twice in DSG3. Desmogleins are therefore considerably larger than desmocollins; for example mature human DSG1 has a molecular weight of 107,578[12] compared with 84,633 and 78,447 for the two splice variants of human DSC2.[8]

A particularly striking difference between both types of desmosomal cadherins and the classic cadherins is the high content of cysteine in the former, specifically in the cytoplasmic domain, with 8 in this domain in human DSC2a, 4 in DSC2b, 10 in DSG1, but none in the cadherins. Presumably these cysteines play an important role in the molecular interactions of the desmosomal cadherins with other plaque proteins. Indeed, there is evidence that plakoglobin interacts with desmogleins via their cysteine-rich cytoplasmic tails both from studies with pemphigus autoantisera[15] and from sulphydryl crosslinking experiments (A.I. Magee, unpublished data).

Both the desmocollins and desmogleins occur as a number of different isoforms coded by individual genes and expressed in cell type-specific patterns. For some time it has been known that certain anti-desmocollin[16] and anti-desmoglein[17] antibodies only recognize desmosomes in epidermal suprabasal cells. While it was possible that these results were due to differential masking of epitopes, it did suggest that there could be variation in the composition of these junctions. In the case of the desmocollins, amino acid sequencing of proteolytic products suggested that there were different isoforms of the desmocollins which were expressed in different locations in human epidermis.[18] Moreover, the first published sequences of desmocollins from bovine snout epithelium[5-7] and from human keratinocytes[8] showed only 50% identity of their deduced amino acid sequences. Species orthologues of other cadherin subtypes show substantially greater identity than this, so the possibility existed that these desmocollins were not species orthologues but represented different subfamilies. This has now been confirmed for three bovine and human desmocollins, termed desmocollins types 1, 2 and 3[19-23] (and I.A. King, unpublished data). In the case of the desmogleins, determinations of amino acid sequences derived from human cDNA clones encoding desmoglein from different kinds of cells[24] or selected using autoantibodies from the blistering skin disease pemphigus vulgaris[25] has also revealed the existence of cell type-specific isoforms for these proteins distinct from the originally described human desmoglein[12-14] (see refs. 26 and 27 for reviews). The desmocollin and desmoglein proteins are abbreviated to Dsc1, Dsg1 etc. whilst the genes are termed DSC1, DSG1 etc.[28]

By comparing the translated amino acid sequences of complete or nearly complete cDNA clones, the relationship between the desmocollin clones isolated from human, bovine, murine and canine sources is now clear (Table 1.1); thus the originally isolated human clone (DGII/III)[8] is orthologous with a type 2 bovine clone,[19] with a murine clone isolated from an 8.5 day embryo cDNA library,[29] and with a canine clone from MDCK cells (G.A. Roberts, S.C. Pidsley and R.S. Buxton, unpublished data) but having a distinct sequence from type 1 desmocollin (DSC1)[5-7,20,21] and from a further bovine[23] and human (I.A. King, unpublished data) desmocollin isoform (type 3) recently isolated. This relationship differs from that recently described[23] where it was not realized that type 2 bovine desmocollin (DSC2)[19] was orthologous with the original human desmocollin (DGII/III),[8] previously termed DSC3,[28] and where the new bovine type 3 desmocollin was incorrectly thought to be the orthologue of human DGII/III. In this type of comparison it is essential that complete rather than partial protein sequences are compared. In this chapter the originally isolated human desmocollin (DGII/III)[8]

Table 1.1. Comparison of the amino acid sequences derived from type 'a' desmocollin cDNA clones from various species

Type	CAR	Ref. to cDNA cloning	EMBL Accession no. for cDNA	Tissue distribution	Expression layer in epidermis	HDSC1	HDSC2	HDSC3	BDSC1	BDSC2	BDSC3	MDSC2	CDSC2
HDSC1	YAT	20-22	X72925	epidermis only[20]	suprabasal[20,33]	**100%**							
HDSC2	FAT	8	X56807	various epithelia,[20] myocardium[e]	basal[20, 33]	55%	**100%**						
HDSC3	YAS	a	a	ND	basal and suprabasal[a]	51%	67%	**100%**					
BDSC1	YAT	6, 7	X56966	epidermis only[19]	very suprabasal [23]	**81%**	53%	52%	**100%**				
BDSC2	FAT	19	M81190	various epithelia myocardium[19,23]	suprabasal[23]	54%	**78%**	64%	55%	**100%**			
BDSC3	YAS	23, b	X75783	various stratified epithelia[23]	basal[23]	53%	67%	**82%**	55%	65%	**100%**		
MDSC2	FAT	29, c	X73885 L33779	various epithelia[c]	suprabasal[c]	54%	**75%**	62%	52%	**72%**	66%	**100%**	
CDSC2	FAT	d	d	ND	ND	55%	**83%**	64%	55%	**79%**	68%	**74%**	**100%**

Intraspecies and interspecies comparisons:		Interspecies comparisons:	
Type 1 to type 2	53% - 55%	Type 1 to type 1	81%
Type 1 to type 3	51% - 55%	Type 2 to type 2	72% - 83%
Type 2 to type 3	62% - 68%	Type 3 to type 3	82%

Figures given are percentage identities using the GAP computer program. The sequences used are either full-length cDNAs or nearly full-length extending beyond the CAR site. Identical or orthologous percentage identities are bold. H is human, B is bovine and M is murine. CAR is the potential cell adhesion recognition site. ND = not determined.

[a] I.A. King, unpublished data.

[b] K.K.M. Yue, J.L. Holton, J.P. Clarke, J.L.M. Hyam, T. Hashimoto, M.A.J. Chidgey and D.R. Garrod, personal communication.

[c] J.E. Lorimer, L.S. Hall, J.P. Clarke, J.E. Collins, T.P. Fleming and D.R. Garrod, personal communication.

[d] G.A. Roberts, S.C. Pidsley and R.S. Buxton, unpublished data.

[e] B.D. Angst, R.S. Buxton and A.I. Magee, unpublished data.

and the mouse desmocollin[29] are therefore renamed DSC2 so that there is a standard nomenclature between the isoforms of different species. However, the reported expression patterns of the orthologous isoforms in human and bovine appear not to be identical (see below). The relationship between the human and bovine isoforms is detailed in Table 1.3 (this is a revised version of the table which appeared in ref. 28). Note that types 2 and 3 have more similar sequences than either with type 1. It appears that the potential cell adhesion recognition site (CAR), which is a sequence in the first extracellular domain, by comparison with the classical cadherins,[30-32] is conserved amongst desmocollins of the same type between different species and might be the best indicator of type at the present time. The significance of the CAR in the desmosomal cadherins has not yet been tested.

The range of species from which the desmoglein cDNAs have been isolated is less extensive, there being only a bovine and a partial murine DSG1 besides the human DSG1, 2 and 3 clones (Table 1.2). The CAR of human and bovine DSG1 and human DSG3 (both suprabasally expressed) is conserved as RAL, whilst it is YAL in the basally expressed human DSG2. In general the desmogleins are more dissimilar between types than the desmocollins, partly at least because of different numbers of the cytoplasmic repeat sequence in the various desmoglein isoforms.

The expression of these different desmocollin and desmoglein isoforms is related, in epidermis, to the differentiation of the keratinocytes; thus in human epidermis (foreskin) DSC2 and DSG2 are basally expressed, DSG3 is expressed immediately suprabasally whilst DSC1 and

Table 1.2. Comparison of the amino acid sequences derived from desmoglein cDNA clones from various species

Type	CAR	Ref. to cDNA cloning	EMBL Accession no. for cDNA	Tissue distribution	Expression layer in epidermis	HDSG1	HDSG2	HDSG3	BDSG1	MDSG1
HDSG1	RAL	12, 14	X56654	epidermis only[36]	very suprabasal[33]	**100%**				
HDSG2	YAL	24, 36	Z26317	various epithelia, myocardium[36,a]	basal[33]	40.0%	**100%**			
HDSG3	RAL	25	M76482	epidermis only[36]	suprabasal[33]	48.4%	39.4%	**100%**		
BDSG1	RAL	10, 24	X58466	ND	ND	**80.8%**	38.0%	49.8%	**100%**	
MDSG1		29	X74335	ND	ND					**100%**

Figures given are percentage identities using the GAP computer program. The sequences used are either full-length cDNAs or nearly full-length extending beyond the CAR site. Identical or orthologous percentage identities are bold. H is human, B is bovine and M is murine. CAR is the potential cell adhesion recognition site. ND = not determined. The MDSG1 clone is only partial and is not included in the comparisons. The clone does not extend as far as the potential CAR.
[a] B.D. Angst, R.S. Buxton and A.I. Magee, unpublished data.

Table 1.3. Gene and protein names of the desmosomal cadherins

Gene name	Proteins	Human synonyms	Ref. to human cDNA cloning	Bovine equivalent	Ref. to bovine cDNA cloning	Human chromosomal assignment
Desmocollins						
Dsc1	Dsc1aDSC1b	DGIV/DGV	20, 21	band 4a[92, 93] Type 1 desmocollin[19] band 4b[92, 93] desmoglein II[94]	5-7	18[21]
Dsc2	Dsc2aDsc2b	DGII/DGIII	8	Type 2 desmocollin[19]	19	18[21]
Dsc3	Dsc3aDsc3b		a	Type 3 desmocollin[23]	23	18[b]
Desmogleins						
Dsg1	Dsg1	DGI	12, 14	band 3[92, 93], desmoglein I[94], 150-165K[95]	10, 11	18[46]
Dsg2	Dsg2	HDGC	24, 36			18[64]
Dsg3	Dsg3	PVA	25			18[65]

[a] I.A. King, unpublished data.
[b] R.S.Buxton, unpublished data.

DSG1 are expressed more suprabasally.[33] Another report also describes expression of DSC2 in human forehead epidermis as being in several basal layers, while DSC1 was restricted to the suprabasal layers.[20] Preliminary observations suggest that human DSC3 is expressed throughout the lower half of foreskin epidermis including the basal layer (I.A. King and K.H. Sullivan, unpublished data). In contrast, expression of DSC2 in bovine nasal epidermis has been reported to be suprabasal being weak or absent from the basal layer, whilst DSC3 was strongly basal extending into the first few suprabasal layers.[23] The reasons for these apparent differences in expression between human and bovine epidermis are not clear and may reflect differences in interpretation. It is of course still possible but perhaps unlikely that we have not yet exhausted the range of isoforms and are still not comparing like with like.

Two of the three desmogleins are the autoantigens of the autoimmune blistering skin diseases pemphigus vulgaris (DSG3)[25] and pemphigus foliaceous (DSG1),[34,35] where there is acantholysis due to the loss of cell adhesion between keratinocytes (see ref. 1 for a review). The position of acantholysis in these diseases correlates with the localization of the gene transcripts as demonstrated by in situ hybridization.[33]

It is an interesting possibility that the pairs of

desmogleins and desmocollins expressed in the various layers of the epidermis are obligatory partners and have co-evolved to carry out subtly different functions.

From expression pattern studies, it appears that the desmocollin DSC2 and the desmoglein DSG2 are widely expressed in many simple and stratified epithelia, whereas DSC1, DSG1 and DSG3 are restricted to certain specialized epithelia.[19,20,23,36] Thus the desmocollin and desmoglein isoforms which are basally expressed in epidermis[33] appear to be the most widely distributed and, at least for DSC2, are synthesized in the 8.5 day[29] and mid 16-cell stage mouse embryos (J.E. Collins, J.E. Lorimer, D.R. Garrod, S.C. Pidsley, R.S. Buxton and T.P. Fleming, unpublished data).

The Desmoplakin Family

cDNAs from three members of this family have been isolated, desmoplakin (DP),[37,38] a prominent component of the desmosomal plaque, the bullous pemphigoid antigen (BPAG1),[39-43] present in the plaque of hemidesmosomes and the intermediate filament-associated protein plectin (PL).[44] Structural analyses of their sequences have indicated that these proteins probably exist as dumbbell-shaped dimers with a central rod and globular end domains (for review see ref. 45). BPAG1 is more closely related to PL in the N-terminal domain, whereas in the C-terminal domain, containing a 38-residue repeating motif, DP and PL are closer in sequence.

Quantitatively the most abundant desmosomal plaque proteins are the desmoplakins (DP). A single gene in the human (DSP) encodes at least two polypeptides (DPI, M_r = 310 kD; DPII, M_r = 240 kD) which are generated by translation of alternatively spliced mRNAs.[37,38,46] Both variants are expressed in all desmosome-bearing cells, at various ratios, with the exception of heart which lacks DPII.[47] Additionally, DP is found as a component of a novel junction, the syndesmos, which does not contain any other known desmosomal components.[48] DPI and II share N- and C-terminal domains, but DPI contains a 1797 bp insertion which encodes a coiled-coil rod domain ~ 130 nm in length. The carboxy-terminal region of DPI and II are predicted to fold into three related globular domains. The rod domain of DPI has a heptad repeat typical of α-fibrous proteins, and has a periodicity of acidic and basic residues which suggests that it may self-aggregate into filamentous structures. The periodicity of acidic and basic residues in the tail is the same as that in the 1B rod domain of intermediate filament (IF) proteins, suggesting the possibility of direct interaction between DP and IF. Evidence consistent with such an interaction has recently been reported.[49] Autoantibodies against the desmoplakins have been found in paraneoplastic pemphigus.[50]

Plectin is a 466 kD polypeptide associated with the sub-plasmalemmal network as well as desmosomes in a wide range of cells.[44] Plectin is known to interact directly with several IF proteins. Sequencing of cDNAs encoding plectin reveals a structure which forms an α-helical coiled coil similar to DP.[44] BPAG1 is one of the protein components of the hemidesmosome, involved in epithelial-basement membrane binding, which is recognized by auto-antibodies from patients with the blistering skin disease bullous pemphigoid. This 230 kD protein is localized to the plaque in the region where IF insert, in an analogous fashion to DP. cDNA cloning again demonstrates a molecule homologous to DP with a coiled-coil rod domain and a similar carboxy-terminal globular domain.[42,43] Although one report suggests that the protein may be partly extracellular and glycosylated,[42] a close examination of the sequence favors the interpretation that it is entirely cytoplasmic. It seems highly likely, therefore, that each of these proteins mediates binding of IF to specific sub-plasmalemmal networks.

The Plakoglobin/Armadillo Family

Plakoglobin is an ~ 83 kD obligatory component of the desmosomal plaque encoded apparently by a single gene producing only one translated product. cDNAs have been sequenced from human,[51] mouse[52] and *Xenopus*.[53] The protein consists of unique N- and C-terminal domains and a central domain containing 12-13 imperfect repeats of a 42 amino acid sequence. Multiple copies of a similar motif have been found in the product of the *Drosophila* segment polarity gene *armadillo* and are hence called armadillo (arm) repeats. Sequence comparisons have revealed that a growing number of proteins contain variable numbers of arm repeats.[54] These include the adherens junction protein β-catenin (probably the mammalian homologue of *Drosophila* Armadillo), the tumor suppressor gene product APC, a tyrosine kinase substrate p120, a regulator of nucleotide exchange on small GTP-binding proteins (smgGDS) and most recently another stratification-specific desmosomal plaque protein Band 6 protein (B6P).[55] These proteins can be further grouped together on the basis of sequence homology; Armadillo, β-catenin and plakoglobin forming one group, and B6P and p120 another. The role of the arm repeat is unclear but it may be involved in protein-protein interactions since the arm repeats of β-catenin appear to mediate binding to the APC gene product. Although APC itself contains arm repeats these do not seem to play a role in binding to β-catenin, which instead involves another repeated motif of 15-20 amino acids.[56]

The related molecules plakoglobin and β-catenin may play a similar role in the desmosome and adherens junctions respectively, i.e. that of connecting the plaque to the intermediate filament or actomyosin cytoskeletal systems respectively via linker molecules (desmoplakin and α-catenin). B6P, on the other hand, can interact directly with intermediate filaments,[55] which may contribute in a stratification-specific way to desmosomal plaque-filament binding. In addition to binding to desmosomes plakoglobin can also associate more weakly with cadherin cytoplasmic domains,[57,58] although the role of this interaction is unclear.

There has been considerable interest in these findings for at least two reasons. Firstly, Armadillo is known to function in the *Drosophila* Wingless signal transduction pathway which specifies segment polarity. Secondly, mutations of the APC gene product contribute to oncogenic transformation of cells. These observations imply that molecules containing arm repeats, and particularly those involved in interactions with cadherin-based cell adhesion systems, play key roles in intracellular signaling events controlling cell differentiation and proliferation. Given this role it is interesting that two other members of the family, SmgGDS and p120, are known to be involved in interactions with *bona fide* signaling molecules namely small GTP-binding proteins and tyrosine kinases.

Evidence has accumulated that the mammalian homologue of Wingless, namely Wnt-1, also delivers morphogenetic information via a pathway that involves the arm repeat proteins plakoglobin and β-catenin and cadherins.[59,60]

Other Desmosomal and Hemidesmosomal Genes

The BPAG2 gene, coding for another protein found in the hemidesmosome and the upper lamina, has been cloned by screening expression libraries with bullous pemphigoid autoantibodies[61-63] and is quite unlike any other junction proteins in being a transmembrane collagenous molecule. It is a protein with a molecular weight of 180 kD which has an extracellular C-terminal region containing 15 collagen domains separated by short noncollagenous regions, a central transmembrane sequence and an intracellular N-terminal domain. Results suggest that the major antigenic site recognized by BP-herpes gestationis autoantibodies, from a subepidermal blistering disorder typically occurring during the third trimester of pregnancy, is located between the transmembrane and collagenous domains.[62]

MAPPING OF DESMOSOMAL GENES

To date four loci have been identified on the human genome to which desmosomal and hemidesmosomal genes have been located. These are the desmosomal cadherin locus on chromosome 18, the desmoplakin locus on chromosome 6, plakoglobin on chromosome 7 and bullous pemphigoid 180 antigen (BPAG2) on chromosome 10.

All of the known desmosomal cadherin genes, the three desmoglein genes (DSG1, 2 and 3) and the three desmocollin genes (DSC1, 2 and 3), have been located on human chromosome 18 by PCR analysis on somatic cell hybrids[21,29,46,64,65] (and R.S. Buxton, unpublished data for DSC3). Clustering of all of the desmosomal cadherin genes so far examined has been observed at 18q12.1; thus linkage between DSG1 and DSG2 has been reported at 18q12.1 in a deletion panel of somatic cell hybrids,[66] DSG1 and DSG3 have been mapped by FISH (fluorescent in situ hybridization) to band q12 on chromosome 18 and co-localized on a 320 kb genomic fragment,[67] and close linkage between a desmoglein (DSG1) and a desmocollin gene (DSC2) on mouse chromosome 18, in a region having conserved synteny with human chromosome 18, has been demonstrated by interspecific backcross analysis.[29] Of the genes coding for other members of the cadherin superfamily not found in desmosomes, which are expressed in different tissues, only N-cadherin maps to chromosome 18;[68] E-cadherin is located on human chromosome 16[69] and mouse chromosome 8,[70] and P-cadherin is also on mouse chromosome 8.[71]

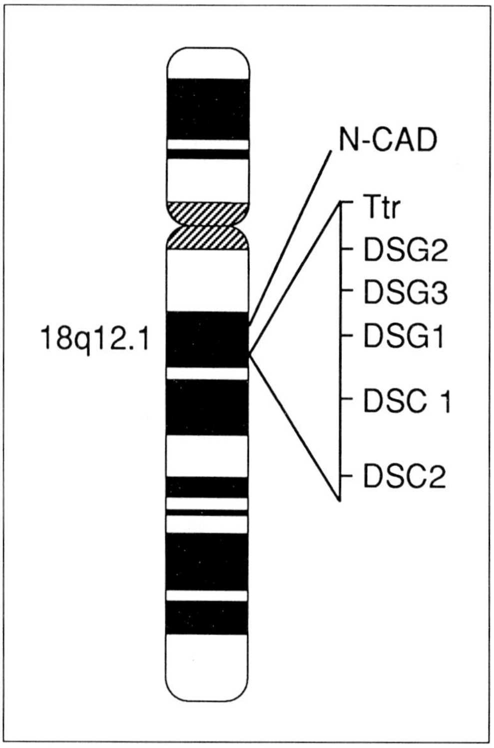

Fig. 1.2. Map of the desmosomal cadherin gene cluster on human chromosome 18. The map is not drawn to scale; it shows the order of the desmosomal cadherin genes in relation to the Ttr (Palb) gene. The orientation of this cluster with relation to the centromere is not known (D. Simrak, C.M.E. Cowley, R.S. Buxton and J. Arnemann, unpublished data).

Recent analysis of yeast artificial chromosome clones (D. Simrak, C.M.E. Cowley, R.S. Buxton and J. Arnemann, unpublished data) has demonstrated that the desmoglein and desmocollin genes are very closely linked, and at least for the desmogleins there is spatial colinearity between the order of DSG genes and their expression within the epidermis, reminiscent of the ordered expression of the hox genes within the developing embryo.[72] Thus the order of genes is DSG2-DSG3-DSG1, the same as order of expression from the basal to the suprabasal layer. This raises the intriguing possibility that the organization of the desmoglein gene cluster is required for properly regulated gene expression. The DSC genes are very closely linked and a YAC contig joining the DSG and DSC genes has now been isolated. The DSC1 gene is closest to the DSG cluster with DSC2 further away; the position of DSC3 is not yet known. A map of the desmosomal cadherin genes on human chromosome 18 is presented in Figure 1.2. Another member of the cadherin superfamily, N-cadherin, although linked on chromosome 18, appears to be some distance away (J. Arnemann, unpublished data).

Conservation of intron-exon boundaries between bovine DSG1 and a classical cadherin, mouse E-cadherin,[73] has been described,[74] supporting the contention that the desmosomal cadherins and the classical cadherins have evolved by duplication of an ancestral gene and subsequent evolutionary divergence.

Two genes of the desmoplakin family have been mapped to chromosome 6: BPAG1 coding for the bullous pemphigoid antigen[40] by Southern blotting to a panel of somatic cell hybrids and by fluorescent in situ hybridization to 6p11-6p12, and DSP coding for desmoplakin by PCR analysis of somatic cell hybrids, suggesting a location at 6p21-ter.[46] A more detailed analysis of these loci has not yet been reported. Of the plakoglobin family, only plakoglobin itself (JUP) has been mapped, by PCR analysis of somatic cell hybrids to chromosome 7.[46] The gene coding for the BP180 protein, BPAG2, has been mapped to human chromosome 10q24.3 by chromosomal in situ hybridization.[75]

MUTATIONS AND POLYMORPHISMS

At present no inherited diseases have been demonstrated to be due to mutations in desmosomal genes; however two likely candidates from histological studies are the autosomal dominant skin diseases Darier's disease and Hailey-Hailey disease. In the former for example, there appears to be an abnormality in the desmosome-keratin filament complex, leading to a breakdown in cell adhesion.[76] Failure of the mechanisms regulating the formation or breakdown of this type of junction could account for the features of Darier's disease and on this basis a number of molecules would seem to be likely candidates. These would include the various keratin isoforms which have been mapped to chromosomes 12q and 17q, the desmocollin and desmoglein isoforms on chromosome 18q, perhaps E-cadherin, mapping to chromosome 16q, and especially the proteins of the desmosomal plaque thought to make the link with the keratins, viz. desmoplakin mapping on chromosome 6p and plakoglobin on chromosome 7. In fact immunohistochemical studies have indicated that the keratins are normal in perilesional skin of Darier's patients,[77] but there was an apparent delay in the expression of the suprabasal skin specific keratins. A similar study with antibodies against desmosomal components detected some reduced expression in lesional skin but it was concluded that such changes were probably not primary deficiencies in these proteins.[78] Another type of mechanism which could give rise to the observed phenotype is protease activation which could mediate the acantholysis seen in Darier's disease, but the changes observed[79] may of course be secondary to other alterations. Recent results from linkage analysis of two large pedigrees have shown definite linkage to chromosome 12q,[80-82] the chromosome where some of the keratin genes, predominantly type II (neutral-basic) except for keratin 18, are located. However, the keratins located on 12q have been reported to map at some distance proximal to the linkage reported for Darier's disease. These results give a location of Darier's disease at 12q23-q24.1, more distal than the known keratin cluster at 12q11-q13, so it appears that this is not a mutation in the known keratin cluster and other candidate genes must be sought. Possible candidate genes which have not yet been mapped include band 6, a member of the plakoglobin/armadillo family[83,84] which, since it is found in desmosomes of stratified tissue, is a particularly attractive candidate for Darier's and Hailey-Hailey diseases. Less likely perhaps since it is also found in hemidesmosomes, is IFAP 300, a probable linker of intermediate filaments to desmosomes and hemidesmosomes.[85] Hailey-Hailey disease (benign familial pemphigus) has clinical features which overlap with those of Darier's disease[86] and could be allelic or be a mutation in a related gene. Evidence has recently been obtained that Hailey-Hailey disease is not linked to the Darier's disease gene (T. Strachan, personal communication) nor linked to any known desmosomal gene (E. Epstein, personal communication).

One possible mouse mutation which could have a defect in the desmosomal genes is the bal (balding) mutation which has been found to map in the same region of mouse chromosome 18 as DSG1 and DSC2.[87] The bal mutation gives rise to a hair loss phenotype, and because the desmosomal cadherins are expressed in the epidermis they are potential candidate genes for this mutation.

Currently RFLPs which could be used to determine linkage to human inherited diseases have been described for the desmosomal cadherins DSG1[88] and DSC2,[89] and mouse interspecific RFLPs have been described for the same genes.[29] RFLPs have also been described for BPAG1 and BPAG2 and used to exclude these genes as candidates for dominant epidermolysis bullosa simplex.[90]

An interesting sequence variation in the fifth extracellular domain of bovine DSG1 has been reported which appears to be a polymorphism rather than a case of alternative splicing.[74] This is a region harboring epitopes recognized by adhesion-disrupting antibodies, but the large degree of sequence difference between the polymorphic variants confers doubt on the importance of this region in cell adhesion, suggesting that it may act as a spacer or hinge region.

REGULATION OF GENE EXPRESSION

The controls which regulate the expression of the different desmosomal cadherin isoforms with reference to, for example the layer of keratinocyte differentiation[46] or to the stage of the developing embryo (J.E. Collins, J.E. Lorimer, D.R. Garrod, S.C. Pidsley, R.S. Buxton and T.P. Fleming, unpublished data) are not currently understood to the same extent as those affecting the keratin genes. Cloning of genomic sequences has shown that there are consensus sequences for various transcription factors in the 5' upstream regions of DSC2 (M.D. Marsden and R.S. Buxton, unpublished data), DSG3[91] and DSG1, and has indicated considerable homology between the 5' regions of the two suprabasally expressed DSG genes (M.J. Adams and R.S. Buxton, unpublished data).

CONCLUDING REMARKS

The story of desmosomal genes is one of genes which have become duplicated and have subsequently evolved to fulfill various niches, thus generating a great complexity and subtlety of junction structure and function. The structure and regulation of expression of these genes is clearly an area of great interest which may, particularly in the case of the desmosomal cadherins, involve questions of genomic organization, and could have relevance for such conditions as psoriasis, where epidermal gene expression gets out of control. Whether the close linkage and gene order of the desmosomal cadherin genes really has relevance for gene expression or whether it is merely a consequence of a recent evolutionary event remains to be seen. The close linkage of the desmocollin and desmoglein genes could suggest that they are expressed as pairs in desmosomes; they could, for example be partners in heterophilic cell adhesion.

ACKNOWLEDGMENTS

We should like to thank Dai Rees for his interest and support, and the other members of our laboratories for their help and for their unpublished results, especially Michael Adams, Katie Cowley, Mark Marsden, Sara Pidsley, Gareth Roberts, Danijela Simrak, and Kate Sullivan. Our work was supported by the Medical Research Council in London and by the Deutsche Forschungsgemeinschaft (DFG; Ar-224/1) in Frankfurt.

REFERENCES

1. Stanley JR. Cell adhesion molecules as targets of autoantibodies in pemphigus and pemphigoid, bullous diseases due to defective epidermal cell adhesion. Advan Immunol 1993; 53:291-325.
2. Penn EJ, Hobson C, Rees DA, et al. Structure and assembly of desmosome junctions: biosynthesis, processing, and transport of the

major protein and glycoprotein components in cultured epithelial cells. J Cell Biol 1987; 105:57-68.

3. Parrish EP, Marston JE, Mattey DL, et al. Size heterogeneity, phosphorylation and transmembrane organisation of desmosomal glycoproteins 2 and 3 (desmocollins) in MDCK cells. J Cell Sci 1990; 96:239-248.
4. Cowin P, Mattey D, Garrod D. Identification of desmosomal surface components (desmocollins) and inhibition of desmosome formation by specific Fab'. J Cell Sci 1984; 70:41-60.
5. Collins JE, Legan PK, Kenny TP, et al. Cloning and sequence analysis of desmosomal glycoprotein 2 and glycoprotein 3 (desmocollins): cadherin-like desmosomal adhesion molecules with heterogeneous cytoplasmic domains. J Cell Biol 1991; 113:381-391.
6. Koch PJ, Goldschmidt MD, Walsh MJ, et al. Amino acid sequence of bovine muzzle epithelial desmocollin derived from cloned cDNA: a novel subtype of desmosomal cadherins. Differentiation 1991; 47:29-36.
7. Mechanic S, Raynor K, Hill JE, et al. Desmocollins form a subset of the cadherin family of cell adhesion molecules. Proc Natl Acad Sci USA 1991; 88:4476-4480.
8. Parker AE, Wheeler GN, Arnemann J, et al. Desmosomal glycoproteins II and III: cadherin-like junctional molecules generated by alternative splicing. J Biol Chem 1991; 266:10438-10445.
9. Troyanovsky SM, Eshkind LG, Troyanovsky RB, et al. Contributions of cytoplasmic domains of desmosomal cadherins to desmosome assembly and intermediate filament anchorage. Cell 1993; 72:561-574.
10. Koch PJ, Walsh MJ, Schmelz M, et al. Identification of desmoglein, a constitutive desmosomal glycoprotein, as a member of the cadherin family of cell adhesion molecules. Eur J Cell Biol 1990; 53:1-12.
11. Goodwin L, Hill JE, Raynor K, et al. Desmoglein shows extensive homology to the cadherin family of cell adhesion molecules. Biochem Biophys Res Commun 1990; 173:1224-1230.
12. Wheeler GN, Parker AE, Thomas CL, et al. Desmosomal glycoprotein I, a component of intercellular desmosome junctions, is related to the cadherin family of cell adhesion molecules. Proc Natl Acad Sci USA 1991; 88:4796-4800.
13. Wheeler GN, Buxton RS, Parker AE, et al. Desmosomal glycoproteins I, II and III: novel members of the cadherin superfamily. Biochem Soc Trans 1991; 19:1060-1064.
14. Nilles LA, Parry DAD, Powers EE, et al. Structural analysis and expression of human desmoglein: a cadherin-like component of the desmosome. J Cell Sci 1991; 99:809-821.
15. Korman NJ, Eyre RW, Klaus-Kovtun V, et al. Demonstration of an adhering-junction molecule (plakoglobin) in the autoantigens of pemphigus foliaceus and pemphigus vulgaris. N Engl J Med 1989; 321:631-635.
16. Parrish EP, Garrod DR, Mattey DL, et al. Mouse antisera specific for desmosomal adhesion molecules of suprabasal skin cells, meninges, and meningioma. Proc Natl Acad Sci USA 1986; 83:2657-2661.
17. Jones JC, Vikstrom KL, Goldman RD. Evidence for heterogeneity in the 160/165 x 10^3 M_r glycoprotein components of desmosomes. J Cell Sci 1987; 88:513-520.
18. King IA, Magee AI, Rees DA, et al. Keratinization is associated with the expression of a new protein related to the desmosomal cadherins DGII/III. FEBS Lett 1991; 286:9-12.
19. Koch PJ, Goldschmidt MD, Zimbelmann R, et al. Complexity and expression patterns of the desmosomal cadherins. Proc Natl Acad Sci USA 1992; 89:353-357.
20. Theis DG, Koch PJ, Franke WW. Differential synthesis of type 1 and type 2 desmocollin mRNAs in human stratified epithelia. Int J Dev Biol 1993; 37:101-110.
21. King IA, Arnemann J, Spurr NK, et al. Cloning of the cDNA (DSC1) coding for human type 1 desmocollin and its assignment to chromosome 18. Genomics 1993; 18:185-194.
22. King IA. Identification of the ATG initiation codon in human type I desmocollin. J Invest Dermatol 1994; 102:822.
23. Legan PK, Yue KKM, Chidgey MAJ, et al. The bovine desmocollin family: a new gene

and expression patterns reflecting epithelial cell proliferation and differentiation. J Cell Biol 1994; 126:507-518.

24. Koch PJ, Goldschmidt MD, Walsh MJ, et al. Complete amino acid sequence of the epidermal desmoglein precursor polypeptide and identification of a second type of desmoglein gene. Eur J Cell Biol 1991; 55:200-208.
25. Amagai M, Klaus-Kovtun V, Stanley JR. Autoantibodies against a novel epithelial cadherin in pemphigus vulgaris, a disease of cell adhesion. Cell 1991; 67:869-877.
26. Magee AI, Buxton RS. Transmembrane molecular assemblies regulated by the greater cadherin family. Curr Opin Cell Biol 1991; 3:854-861.
27. Buxton RS, Magee AI. Structure and interactions of desmosomal and other cadherins. Semin Cell Biol 1992; 3:157-167.
28. Buxton RS, Cowin P, Franke WW, et al. Nomenclature of the desmosomal cadherins. J Cell Biol 1993; 121:481-483.
29. Buxton RS, Wheeler GN, Pidsley SC, et al. Mouse desmocollin (DSC3) and desmoglein (DSG1) genes are closely linked in the proximal region of chromosome 18. Genomics 1994; 21:510-516.
30. Blaschuk OW, Pouliot Y, Holland PC. Identification of a conserved region common to cadherins and influenza strain A hemagglutinins. J Mol Biol 1990; 211:679-682.
31. Blaschuk OW, Sullivan R, David S, et al. Identification of a cadherin cell adhesion recognition sequence. Dev Biol 1990; 139:227-229.
32. Nose A, Tsuji K, Takeichi M. Localization of specificity determining sites in cadherin cell adhesion molecules. Cell 1990; 61:147-155.
33. Arnemann J, Sullivan KH, Magee AI, et al. Stratification-related expression of isoforms of the desmosomal cadherins in human epidermis. J Cell Sci 1993; 104:741-750.
34. Koulu L, Kusumi A, Steinberg MS, et al. Human autoantibodies against a desmosomal core protein in pemphigus foliaceus. J Exp Med 1984; 160:1509-1518.
35. Stanley JR, Koulu L, Klaus-Kovtun V, et al. A monoclonal antibody to the desmosomal glycoprotein desmoglein I binds the same polypeptide as human autoantibodies in pemphigus foliaceus. J Immunol 1986; 136:1227-1230.
36. Schäfer S, Koch PJ, Franke WW. Identification of the ubiquitous human desmoglein, DSG2, and the expression catalogue of the desmoglein subfamily of desmosomal cadherins. Exp Cell Res 1994; 211:391-399.
37. Green KJ, Parry DA, Steinert PM, et al. Structure of the human desmoplakins. Implications for function in the desmosomal plaque. J Biol Chem 1990; 265:2603-2612.
38. Virata MLA, Wagner RM, Parry DAD, et al. Molecular structure of the human desmoplakin I and desmoplakin II amino terminus. Proc Natl Acad Sci USA 1992; 89:544-548.
39. Stanley JR, Tanaka T, Mueller S, et al. Isolation of complementary DNA for bullous pemphigoid antigen by use of patients' autoantibodies. J Clin Invest 1988; 82:1864-1870.
40. Sawamura D, Nomura K, Sugita Y, et al. Bullous pemphigoid antigen (BPAG1): cDNA cloning and mapping of the gene to the short arm of human chromosome 6. Genomics 1990; 8:722-726.
41. Amagai M, Hashimoto T, Tajima S, et al. Partial cDNA cloning of the 230-kD mouse bullous pemphigoid antigen by use of a human monoclonal anti-basement membrane zone antibody. J Invest Dermatol 1990; 95:252-259.
42. Sawamura D, Li KH, Chu ML, et al. Human bullous pemphigoid antigen (BPAG1). Amino acid sequences deduced from cloned cDNAs predict biologically important peptide segments and protein domains. J Biol Chem 1991; 266:17784-17790.
43. Tanaka T, Parry DAD, Klaus-Kovtun V, et al. Comparison of molecularly cloned bullous pemphigoid antigen to desmoplakin-I confirms that they define a new family of cell adhesion junction plaque proteins. J Biol Chem 1991; 266:12555-12559.
44. Wiche G, Becker B, Luber K, et al. Cloning and sequencing of rat plectin indicates a 466-kD polypeptide chain with a three-domain structure based on a central alpha-helical coiled coil. J Cell Biol 1991;

114:83-99.
45. Green KJ, Virata MLA, Elgart GW, et al. Comparative structural analysis of desmoplakin, bullous pemphigoid antigen and plectin: members of a new gene family involved in organization of intermediate filaments. Int J Biol Macromol 1992; 14:145-153.
46. Arnemann J, Spurr NK, Wheeler GN, et al. Chromosomal assignment of the human genes coding for the major proteins of the desmosome junction, desmoglein DGI (DSG), desmocollins DGII/III (DSC), desmoplakins DPI/II (DSP), and plakoglobin DPIII (JUP). Genomics 1991; 10:640-645.
47. Angst BD, Nilles LA, Green KJ. Desmoplakin II expression is not restricted to stratified epithelia. J Cell Sci 1990; 97:247-257.
48. Schmelz M, Franke WW. Complexus adhaerentes, a new group of desmoplakin-containing junctions in endothelial cells: The syndesmos connecting retothelial cells of lymph nodes. Eur J Cell Biol 1993; 61:274-289.
49. Stappenbeck TS, Bornslaeger EA, Corcoran CM, et al. Functional analysis of desmoplakin domains: specification of the interaction with keratin versus vimentin intermediate filament networks. J Cell Biol 1993; 123:691-705.
50. Oursler JR, Labib RS, Arissabdo L, et al. Human autoantibodies against desmoplakins in paraneoplastic pemphigus. J Clin Invest 1992; 89:1775-1782.
51. Franke WW, Goldschmidt MD, Zimbelmann R, et al. Molecular cloning and amino acid sequence of human plakoglobin, the common junctional plaque protein. Proc Natl Acad Sci USA 1989; 86:4027-4031.
52. Butz S, Stappert J, Weissig H, et al. Plakoglobin and beta-catenin: distinct but closely related. Science 1992; 257:1142-1144.
53. Fouquet B, Zimbelmann R, Franke WW. Identification of plakoglobin in oocytes and early embryos of *Xenopus laevis*: maternal expression of a gene encoding a junctional plaque protein. Differentiation 1992; 51:187-194.
54. Peifer M, Berg S, Reynolds AB. A repeating amino acid motif shared by proteins with diverse cellular roles. Cell 1994; 76:789-791.
55. Hatzfeld M, Kristjansson GI, Plessmann U, et al. Band 6 protein, a major constituent of desmosomes from stratified epithelia, is a novel member of the *armadillo* multigene family. J Cell Sci 1994; 107:2259-2270.
56. Peifer M. Cancer, catenins, and cuticle pattern: a complex connection. Science 1993; 262:1667-1668.
57. Piepenhagen PA, Nelson WJ. Defining E-cadherin-associated protein complexes in epithelial cells: plakoglobin, beta- and gamma-catenin are distinct components. J Cell Sci 1993; 104:751-762.
58. Knudsen KA, Wheelock MJ. Plakoglobin, or an 83-kD homologue distinct from beta-catenin, interacts with E-cadherin and N-cadherin. J Cell Biol 1992; 118:671-679.
59. Bradley RS, Cowin P, Brown AMC. Expression of Wnt-1 in PC12 cells results in modulation of plakoglobin and E-cadherin and increased cellular adhesion. J Cell Biol 1993; 123:1857-1865.
60. Hinck L, Nelson WJ, Papkoff J. Wnt-1 modulates cell-cell adhesion in mammalian cells by stabilizing beta-catenin binding to the cell adhesion protein cadherin. J Cell Biol 1994; 124:729-741.
61. Giudice GJ, Squiquera HL, Elias PM, et al. Identification of two collagen domains within the bullous pemphigoid autoantigen, BP180. J Clin Invest 1991; 87:734-738.
62. Giudice GJ, Emery DJ, Diaz LA. Cloning and primary structural analysis of the bullous pemphigoid autoantigen BP180. J Invest Dermatol 1992; 99:243-250.
63. Hopkinson SB, Riddelle KS, Jones JC. Cytoplasmic domain of the 180-kD bullous pemphigoid antigen, a hemidesmosomal component: molecular and cell biologic characterization. J Invest Dermatol 1992; 99:264-270.
64. Arnemann J, Spurr NK, Magee AI, et al. The human gene (DSG2) coding for HDGC, a second member of the desmoglein subfamily of the desmosomal cadherins, is, like DSG1 coding for desmoglein DGI, assigned to chromosome 18. Genomics 1992; 13:484-486.
65. Arnemann J, Spurr NK, Buxton RS. The

human gene (DSG3) coding for the pemphigus vulgaris antigen is, like the genes coding for the other two known desmogleins, assigned to chromosome 18. Hum Genet 1992; 89:347-350.
66. Overhauser J, Mewar R, Rojas K, et al. STS map of genes and anonymous DNA fragments on human chromosome 18 using a panel of somatic cell hybrids. Genomics 1993; 15:387-391.
67. Wang Y, Amagai M, Minoshima S, et al. The human genes for desmogleins (DSG1 and DSG3) are located in a small region on chromosome 18q12. Genomics 1994; 20:492-495.
68. Walsh FS, Barton CH, Putt W, et al. N-cadherin gene maps to human chromosome 18 and is not linked to the E-cadherin gene. J Neurochem 1990; 55:805-812.
69. Mansouri A, Spurr N, Goodfellow PN, et al. Characterization and chromosomal localization of the gene encoding the human cell adhesion molecule uvomorulin. Differentiation 1988; 38:67-71.
70. Eistetter HR, Adolph S, Ringwald M, et al. Chromosomal mapping of the structural gene coding for the mouse cell adhesion molecule uvomorulin. Proc Natl Acad Sci USA 1988; 85:3489-3493.
71. Hatta M, Miyatani S, Copeland NG, et al. Genomic organization and chromosomal mapping of the mouse P-cadherin gene. Nucleic Acids Res 1991; 19:4437-4441.
72. Whiting J, Marshall H, Cook M, et al. Multiple spatially specific enhancers are required to reconstruct the pattern of Hox-2.6 gene expression. Genes & Develop 1991; 5:2048-2059.
73. Ringwald M, Baribault H, Schmidt C, et al. The structure of the gene coding for the mouse cell adhesion molecule uvomorulin. Nucleic Acids Res 1991; 19:6533-6539.
74. Puttagunta S, Mathur M, Cowin P. Structure of DSG1, the bovine desmosomal cadherin gene encoding the pemphigus foliaceus antigen—evidence of polymorphism. J Biol Chem 1994; 269:1949-1955.
75. Li K, Sawamura D, Giudice GJ, et al. Genomic organization of collagenous domains and chromosomal assignment of human 180-kDa bullous pemphigoid antigen-2, a novel collagen of stratified squamous epithelium. J Biol Chem 1991; 266:24064-24069.
76. Caulfield JB, Wilgram GF. An electron microscopic study of dyskeratosis and acantholysis in Darier's disease. J Invest Dermatol 1963; 41:57-65.
77. Burge SM, Fenton DA, Dawber RP, et al. Darier's disease: an immunohistochemical study using monoclonal antibodies to human cytokeratins. Br J Dermatol 1988; 118:629-640.
78. Burge SM, Garrod DR. An immunohistological study of desmosomes in Darier's disease and Hailey-Hailey disease. Br J Dermatol 1991; 124:242-251.
79. Burge SM, Ryan TJ, Cederholm-Williams SA. Darier's disease: an immunohistochemical study using antibodies to proteases. Br J Dermatol 1989; 121:613-621.
80. Craddock N, Dawson E, Burge S, et al. The gene for Darier's disease maps to chromosome 12q23-q24.1. Hum Molec Genet 1993; 2:1941-1943.
81. Bashir R, Munro CS, Mason S, et al. Localisation of a gene for Darier's disease. Hum Molec Genet 1993; 2:1937-1939.
82. Parfitt E, Burge S, Craddock N, et al. The gene for Darier's disease maps between D12S78 and D12S79. Hum Mol Genet 1994; 3:35-38.
83. Kapprell HP, Owaribe K, Franke WW. Identification of a basic protein of Mr 75,000 as an accessory desmosomal plaque protein in stratified and complex epithelia. J Cell Biol 1988; 106:1679-1691.
84. Schäfer S, Troyanovsky SM, Heid HW, et al. Cytoskeletal architecture and epithelial differentiation: molecular determinants of cell interaction and cytoskeletal filament anchorage. C R Acad Sci [III] 1993; 316:1316-1323.
85. Skalli O, Jones JCR, Gagescu R, et al. IFAP-300 is common to desmosomes and hemidesmosomes and is a possible linker of intermediate filaments to these junctions. J Cell Biol 1994; 125:159-170.
86. Burge SM. Hailey-Hailey disease: the clinical features, response to treatment and prognosis. Br J Dermatol 1992; 126:275-282.
87. Davisson MT, Cook SA, Johnson KR, et al. Balding: a new mutation on mouse chro-

mosome 18 causing hair loss and immunological defects. J Hered 1994; 85:134-136.
88. Arnemann J, Spurr NK, Buxton RS. A MspI-RFLP in the C-terminal part of the gene for desmoglein DGI (DSG). Nucleic Acids Res 1991; 19:5100.
89. Arnemann J, Spurr NK, Buxton RS. A TaqI RFLP in the N-terminal part of the gene for desmocollins DGII/III (DSC). Nucleic Acids Res 1991; 19:6343.
90. Ryynänen M, Knowlton RG, Kero M, et al. Bullous pemphigoid antigens (BPAGs): Identification of RFLPs in human BPAG1 and BPAG2, and exclusion as candidate genes in a large kindred with dominant epidermolysis bullosa simplex. Genomics 1991; 11:1025-1029.
91. Silos S, Tamai K, Uitto J. The pemphigus vulgaris antigen (PVA): characterization of the 5' end of the gene and the flanking DNA. J Invest Dermatol 1994; 102:535.
92. Skerrow CJ, Matoltsy AG. Chemical characterization of isolated epidermal desmosomes. J Cell Biol 1974; 63:524-530.
93. Kapprell H-P, Cowin P, Franke WW, et al. Biochemical characterization of desmosomal proteins isolated from bovine muzzle epidermis: amino acid and carbohydrate composition. Eur J Cell Biol 1985; 36:217-229.
94. Giudice GJ, Cohen SM, Patel NH, et al. Immunological comparison of desmosomal components from several bovine tissues. J Cell Biochem 1984; 26:35-45.
95. Cowin P, Garrod DR. Antibodies to epithelial desmosomes show wide tissue and species cross-reactivity. Nature 1983; 302:148-150.

CHAPTER 2

Desmosomes in Differentiation and Development

David R. Garrod, Jane E. Collins

INTRODUCTION

Desmosomes are complex, macromolecular, intercellular junctions that link together the intermediate filament (IF) cytoskeletons of adjacent cells and are believed to provide strong intercellular adhesion. Together, desmosomes and IFs constitute a scaffold-like structure that extends throughout a tissue and is thought to have a mechanical function, providing resistance to shear and abrasion. This view is based on the much greater abundance of desmosome-IF complexes in tissues, such as epidermis, that are subject to continual wear and tear than in tissues that are not, such as the kidney tubule.

Although this is probably correct, it should be stressed that the adhesive role of the desmosome has not been formally demonstrated. Its location at the cell surface and its rather massive structure (in molecular terms) with mirror-image halves contributed by adjacent cells, strongly suggests an adhesive function. Moreover, when desmosomes are isolated, they remain as intact paired-membrane structures.[1,2] Adhesiveness is also suggested by the demonstration that the two major desmosomal glycoproteins, desmoglein and desmocollin, are members of the cadherin family of Ca^{2+}-dependent cell-cell adhesion molecules.[3,4] Despite such evidence, Ca-induced junction formation between MDCK cells in culture, which includes desmosome assembly,[5] can be completely inhibited with antibodies to E-cadherin, the major glycoprotein of the *zonula adherens*.[6] This may suggest that E-cadherin is relatively more important in cell adhesion than desmosomes, at least in the early stages. We shall show, however, that initial adhesion and mature adhesion are substantially different, and

Molecular Biology of Desmosomes and Hemidesmosomes, edited by Jane E. Collins and David R. Garrod.

that in some circumstances cells appear to maintain adhesion through desmosomes alone. The great importance of desmosomes in adhesion of mature tissues is emphasized in the group of diseases called pemphigus (see chapter 7) where autoantibodies to desmosomal glycoproteins cause loss of cell adhesion and epidermal blistering.

The mechanical view of desmosome-IF function is also a very static and limiting one. The epidermis, where desmosomes are most abundant, is a highly dynamic tissue, the differentiated layers being replaced on average every 24 days. This must involve substantial assembly and disassembly of desmosomes. We now know that both desmogleins and desmocollins are represented by at least three distinct isoforms, the products of different genes.[7] All are expressed in epidermis showing distribution patterns dependent on cellular differentiation.[8-10] This suggests that desmosomes are dynamic structures and may play an essential role in epithelial differentiation and stratification.

How is desmosome assembly regulated? In both the early chick and mouse embryos, desmosomes are first expressed in the extra-embryonic regions.[11,12] In the kidney desmosomes appear to be labile during development presumably in order to facilitate changes in cell contact. Desmosome expression may be regulated by external signals, for example, increased expression by TGFβ and the differentiation promoting factors such as dimethylformamide and retinoic acid.[13-15] Nothing is known about the regulatory mechanisms involved in these responses.

There are many situations where desmosome assembly-disassembly needs to be modulated in order to permit cell motility. These include repositioning and remodeling of tissues during embryonic development, movement and stratification during differentiation of adult epithelia, wound healing, and cancer invasion and metastasis. These processes would be facilitated if desmosomal adhesion were to respond rapidly to signals from both inside and outside the cell. We have shown recently that desmosomal adhesion does indeed respond to signals mediated by intercellular contact and protein kinase C (S. Lloyd, I. Wise, T. Fleming and D. Garrod, unpublished data). These observations suggest an entirely novel view of cell adhesion in epithelial tissues, and how it may be rapidly regulated.

That desmosomes can respond to signals raises the possibility that they, like other adhesive structures, may also transduce signals from outside the cell. The potential role of plakoglobin in signal transduction is discussed by Cowin in chapter 4. Furthermore, the interaction of desmosomes with the IF system provides a direct mechanical linkage between the cell surface and the nucleus. It is possible that desmosomes may signal the position and/or state of differentiation of cells within a stratified epithelium, as well as transmitting mechanical signals indicating the stress on a tissue.[10,16]

Thus the desmosome is emerging as a much more variable and dynamic structure than previously thought. Some aspects of its regulation will now be discussed in detail.

DESMOCOLLINS: MOLECULAR BIOLOGY AND EXPRESSION

Desmosomes contain two major glycoproteins, the desmogleins and the desmocollins (see chapter 1). These were first identified in desmosome preparations by Skerrow and Matoltsy.[17] Later, they were demonstrated to be major components of bovine desmosomal cores, paired membrane structures generated from isolated desmosomes by treatment with metrizamide to remove most of the plaque constituents.[2] On this basis, these glycoproteins are believed to represent the major adhesive components of desmosomes and therefore, the major constituents of the desmosomal intercellular cement or "desmoglea."[2] The glycoproteins of the desmoglea consist of three major bands of relative molecular weights 150 kDa, 115 kDa and 100 kDa, sometimes subdivisible into doublets or triplets. On the basis of their enrichment in the desmoglea these three bands were called desmogleins 1, 2 and 3. Similar components were identified in two-dimensional gels by Franke and colleagues.[18] These authors referred to the polypeptide of highest molecular weight as band 3 and the two lower molecular weight polypeptides as 4a and 4b, suggesting a relationship between the latter band based primarily on the similarity of their isoelectric properties. Immunological evidence also suggested a similar relationship. Thus Cowin and Garrod[19] raised polyclonal antibodies against desmosomal polypeptides purified from desmosomal cores by gel electrophoresis while Cohen et al[20] raised monoclonal antibodies to desmosomal cores. These studies indicated a relationship between desmogleins 2 and 3 and distinguished them from desmoglein 1.

Cowin et al[21] prepared univalent F_{ab}' to polyclonal antibodies specific for the desmosomal glycoproteins. The antibodies to desmogleins 2 and 3, when added to cultures at high concentrations, were able to inhibit temporarily the assembly of desmosomes between MDBK cells, while those to desmoglein 1 were not. This provided preliminary additional evidence for the adhesive role of desmogleins 2 and 3. On the basis of these experiments and the above data suggesting a distinction between the two groups of glycoproteins, Cowin et al renamed desmogleins 2 and 3, desmocollins. The name desmoglein was then reserved for the highest molecular weight polypeptide.

Additional evidence for the distinction between the two groups of desmosomal glycoproteins was provided by Kapprell et al[22] by amino acid analysis. The two desmocollin polypeptides showed related amino acid compositions that were distinct from that of desmoglein.

Desmocollins are constitutive components of desmosomes being present in a wide range of different tissues and vertebrate animal species.[19,23] Wherever they occur, they are represented by a pair of glycoproteins of slightly different molecular weight, and in some cell lines and tissues by multiple polypeptide bands.[24] Heterogeneity among desmocollins was also suggested by immunological differences demonstrated between different strata in epidermis with desmocollin-specific polyclonal antibodies.[25] At this stage, the reasons for these multiple bands and immunological differences were not understood. One possibility seemed to be differences in glycosylation.[24,25] Both desmocollins and desmoglein were shown to be calcium binding proteins.[26,27] These calcium binding studies had been prompted by the knowledge that the assembly of desmosomes in tissue culture may be regulated by extracellular calcium concentration.[5,28-30]

The reason for the size heterogeneity of desmocollins was elucidated by Parrish et al[31] using whole cell trypsinization coupled with metabolic labeling with ^{35}S-methionine and immunoprecipitation. They showed that desmocollins exhibited different sized cytoplasmic domains. They proposed a model suggesting that the extracellular domains of the two desmocollin polypeptides were identical in size while the cytoplasmic domains differed by 4 kDa. Almost simultaneously, Holton et al[4] provided evidence supporting the role of desmocollins in calcium-dependent adhesion, by demonstrating their relationship to the cadherin superfamily of calcium-dependent adhesion molecules. At approximately the same time, Koch et al[3] demonstrated that desmoglein is also a cadherin superfamily member.

Cloning and sequencing of desmocollins from bovine and human sources confirmed these initial observations, demonstrating the cadherin homology in the extracellular domains.[32-35] The size heterogeneity of the cytoplasmic domains was also confirmed and shown to be due to alternative splicing of a 46 bp mini exon.[33,34] This exon contains a stop codon and encodes a unique 11 amino acid COOH-terminus. In the absence of the stop codon, the open reading frame is extended by 54 amino acids so that the shorter RNA encodes the larger desmocollin. Alternative splicing in the cytoplasmic domain is so far unique to desmocollins in the cadherin superfamily.

The desmocollin extracellular domain showed about 30% identity to the most closely related members of the cadherin superfamily and the relationship to desmoglein was at approximately the same level. These cloning and sequencing data therefore confirmed the distinction between desmocollins and desmoglein indicated by earlier studies.

Another indication from earlier work was that desmocollins might be more complex than simply two alternatively-spliced polypeptides. Comparison of the sequences of the first bovine and human desmocollins to be fully sequenced gave a further indication of heterogeneity. These sequences showed only a 50% identity at the amino acid level, a very low figure for species homologues of cadherins. This suggested that the two sequences were not species homologues but distinct desmocollin isoforms.

In support of this view King et al[37] obtained partial sequence of a human epidermal protein sequence which seemed to resemble the bovine sequence more closely than the fully described human sequence. The complexity of the desmocollins was subsequently confirmed by Koch et al[38] who published the complete sequence of a second bovine desmocollin that more closely resembled the first human sequence. It is now clear from the studies of Legan et al[10] that there are at least three desmocollin isoforms, the products of different genes, in bovine tissues. These are termed DSC1, 2 and 3.[7] Each is represented by a longer 'a' form and a shorter 'b' form. The bovine isoforms are taken as the standard for nomenclature,[7] and the bovine isoforms will be described first since there has been some confusion about which isoforms from other species are the true homologues of the bovine molecules.

At the amino acid level, the identity between bovine DSC1b and DSC2b is 52%, between DSC1b and DSC3b 54.9% and between DSC2b and DSC3b 64.7%. The isoforms are encoded by messages of different lengths, 5.8 kb for DSC1, 4 to 4.2 kb for DSC2 and 7.2 to 7.5 kb for DSC3,[10,33,38] and show distinct expression patterns in bovine tissues[10] (K. Yue and D. Garrod, unpublished data). DSC2 appears to be ubiquitously expressed in epithelia and cardiac muscle whereas DSC1 and DSC3 are much more restricted. By northern blotting, DSC1 could be detected only in epidermis and tongue, whereas DSC3 was detectable in epidermis, tongue, esophagus and rumen, the latter being a stratified epithelium in the cow. RT-PCR confirmed these distributions for DSC2 and 3, but showed a low level of ubiquitous expression for DSC1.[10] Isoform-specific monoclonal antibodies are available for both DSC1 and DSC3[10] (K. Yue and D. Garrod, unpublished data). Staining with these confirms the tissue distribution indicated by northern blotting. The significance of the low level expression of DSC1 in other tissues is therefore unclear.

The desmocollin isoforms also show strikingly different patterns of expression within stratified epithelia as indicated by in situ hybridization and monoclonal antibody staining. The in situ hybridization studies of Legan et al[10] (Fig. 2.1) showed that DSC1 appears to be associated with terminal differentiation in epidermis and tongue. In epidermis, its expression begins immediately above the basal layer in the dermal papillae of the ridged epithelium of nasal epidermis. At the bottom of the rete ridges however, expression begins five to ten layers of cells from the base. Once initiated, expression appears to continue with equal intensity until it ceases at about the bottom of the granular layer. By contrast, DSC2 is most strongly expressed towards the bases of the rete ridges in approximately the region that shows absence of DSC1. Elsewhere in the epidermis, DSC2 has a lower level of expression that again ceases at about the bottom of the granular layer. It is not possible to decide from in situ hybridization studies whether or not DSC2 is expressed in the basal layer. By contrast, DSC3 is most strongly expressed in the basal layer and its mRNA then declines gradually in the suprabasal layers apparently fading out well below the level of the granular layer. The distribu-

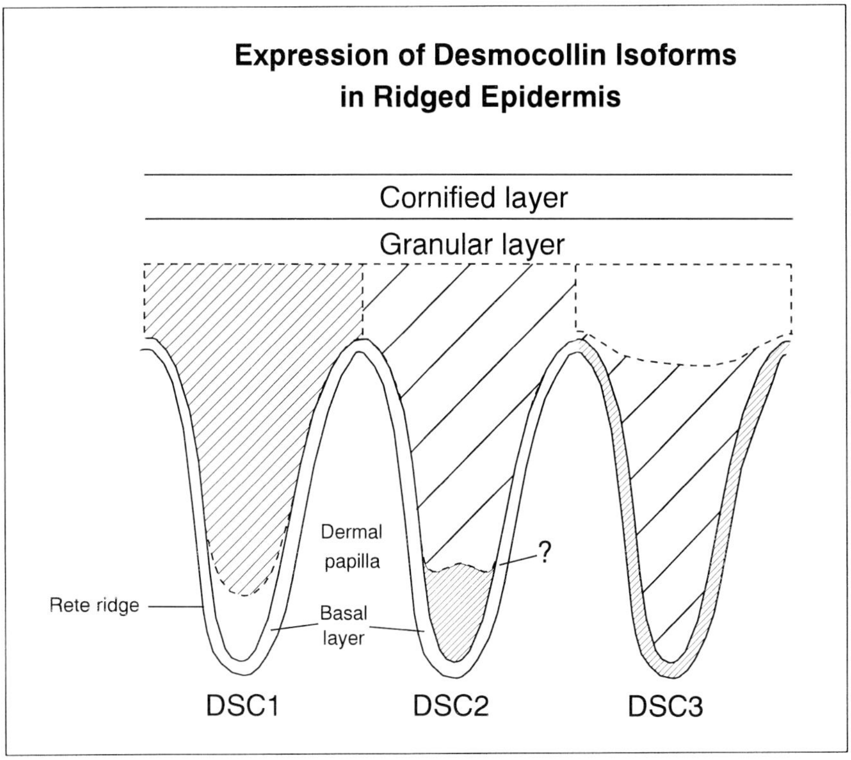

Fig. 2.1. Regions of epidermis where desmocollin gene expression is detectable by in situ hybridization are indicated by cross hatching. Expression is more intense where cross hatching lines are closer together. Basally, the expression of DSC1 and DSC3 glycoproteins, as demonstrated by monoclonal antibody staining corresponds exactly with the pattern of mRNA expression, but apically the glycoprotein expression extends slightly beyond the upper limit of mRNA expression. It is not clear from in situ hybridization whether DSC2 is expressed in the basal layer. In tissue these expression patterns are superimposed so that the expression domains or territories of the three different genes overlap.

tion of the DSC1 and 3 glycoproteins correspond well with the impression obtained from in situ hybridization. At its lower limit the distribution of DSC1 expression appears identical by antibody staining and hybridization whereas in the upper regions, the expression of the protein appears to extend slightly beyond the expression of the mRNA. A similar observation was made for DSC3 where antibody reactivity for DSC3 protein was strongest in the basal layer, faded in the suprabasal layers and extended beyond the upper limit of mRNA expression.

The association of DSC1 with terminal differentiation is confirmed in tongue epithelium where it is associated solely with the posterior region of filiform papillae and the upper suprabasal layers of fungiform papillae[10] (K. Legan and D. Garrod, unpublished data). DSC1 expression is not detectable in other epithelia either by in situ hybridization, northern blotting or antibody staining.

The distributions of DSC2 and DSC3 in the stratified epithelia of tongue, esophagus and rumen are quite characteristic. DSC3 is most strongly expressed in the basal layers whereas DSC2 is strong in the immediately suprabasal layer and several layers above this. In each case, the cessation of mRNA expression appears to cease before the upper limit of desmosome expression as indicated by staining for desmoplakins[10] (K. Yue and D. Garrod, unpublished).

The results obtained for desmocollin isoforms are closely paralleled by the expression patterns of the three desmoglein isoforms, DSG1, DSG2 and DSG3, known from human tissues. Thus Arnemann et al[8] showed that the distribution of DSG mRNAs in human epidermis was related to differentiation, DSG1 being most strongly expressed in the upper suprabasal layers, DSG3 at intermediate levels and DSG2 in the basal layer. Schäfer et al[39] have recently shown that DSG2 is widely expressed in different epithelia and cell lines, whereas DSG1 and DSG3 are largely confined to stratified squamous epithelia.

While the desmocollin and desmoglein patterns are similar, there is one interesting point of difference. Thus, the most widely expressed isoform of desmocollin, DSC2, is most strongly expressed in the suprabasal layers of stratified epithelia, whereas the most widely expressed isoform of desmoglein, DSG2, is most strongly expressed in the basal layer of epidermis.

Two desmocollin isoforms have been cloned and sequenced from human tissues and one from mouse.[9,39,40] There has been some confusion about the nomenclature of these desmocollins. From sequence comparisons, it now appears that the correct nomenclature of the sequences published so far is that shown in Table 2.1. The table shows against each desmocollin isoform the papers that describe its entire or partial cDNA sequence. Thus, for example, DGII/III described by Parker et al[34] should be regarded as DSC2.

Multiple alignment and sequence comparison of the six available desmocollin sequences reveals a number of interesting features (Lorimer et al). Two striking features are that the most highly conserved regions are those towards the NH_2-terminus and the COOH-termini of the alternative cytoplasmic domains. This suggests that these regions are the most functionally important parts of the molecule. It is presumed that the NH_2-terminal region is involved in cell-cell adhesion while the alternatively spliced cytoplasmic termini, being unique among cadherins, has some specific function in relation to desmosomal plaque formation. The experiments of Troyanovsky et al[41] demonstrated such a role for the cytoplasmic region of the 'a' form of DSC2. Chimeras were constructed of the cytoplasmic domains of the 'a' and 'b' forms of desmocollins and desmoglein with the gap junction protein connexin, so that the expressed chimeric protein would be targeted to the cell membrane. It was found that the gap junction-like plaques formed with the COOH-terminal region of the 'a' form supported the development of structures resembling desmosomal plaques with associated plakoglobin, desmoplakin and intermediate filaments. This indicates that the 'a' form cytoplasmic domain contains the molecular signals necessary for plaque assembly and IF association. However, no such structures were formed by the 'b' form chimera nor, for that matter, with desmoglein. Thus, we do not yet have an

Table 2.1. Nomenclature of desmocollins

Correct Designation		Publications Reporting Sequence
Bovine	DSC1	Collins et al;[33] Koch et al;[32] Mechanic et al[35]
"	DSC2	Koch et al[38]
"	DSC3	Legan et al;[10] Troyanovsky et al;[41] K. Yue and D. Garrod, unpublished data.
Human	DSC1	King et al;[95] Theis et al[9]
"	DSC2	Parker et al[34]
Murine	DSC2	Buxton et al;[85] Lorimer et al[86]

Sequence comparisons and multiple alignments between desmocollins can be found in Lorimer et al.

explanation for the function of the 'b' form. On the basis of gel electrophoresis, it appears to be present in an approximately equal amount to the 'a' form in most tissues. Furthermore, preliminary studies with antibodies raised to COOH-terminal peptides of the 'a' and 'b' form show that both forms appear to be present in all cells of the bovine epidermis (D. Garrod, unpublished data). This suggests, but does not prove, that the 'a' and 'b' forms are present together in all desmosomes. We suggest that the 'b' form has some specific function in relation to plaque formation. An interesting difference between the 'a' and 'b' forms is that they were shown to be differentially phosphorylated during desmosome assembly in MDCK cells, the 'a' form being phosphorylated on serine but the 'b' form being unphosphorylated.[31] Phosphorylation presumably functions in regulating desmosomal adhesion.

The differential expression of desmosomal glycoproteins in stratified epithelia strongly suggests that desmosomes play an important role in epidermal differentiation and stratification. The expression of desmocollins in ridge epidermis of the bovine nose showing transition from strongest expression of DSC3 in the basal layer to DSC2 in the immediate suprabasal layers and then to DSC1 in the terminally differentiating region, showed striking similarity to the distribution of stem cells, transit amplifying cells and terminally differentiated cells in ridge epidermis as demonstrated by Lavker and Sun.[42] Thus, a role for desmosomes in signaling the state of differentiation/proliferation of epidermal cells seems possible. One rather surprising feature of desmocollin expression is that the territories in which the different isoforms are expressed substantially overlap each other. This indicates that cells at different levels in the epidermis contain different combinations of the various desmocollin isoforms. Whether these are expressed in the same desmosomes or in distinct junctions is not clear. It seems possible, however, that such a distribution might lead to the existence of a basal-to-apical gradient in desmosomal adhesiveness within a stratified epithelium. Such a gradient might play an important role in the regulation of stratification and differentiation. Loss of function mutations will probably be required to determine the respective functions of the isoforms of the desmosomal glycoproteins.

DESMOSOME EXPRESSION IN MOUSE EARLY EMBRYOS

The study of desmosome formation in the mouse early embryo is of interest from two perspectives. Namely, understanding the role of the desmosome in early development and the study of the molecular regulation of desmosome assembly in a system which is differentiating de novo. The preimplantation embryo develops in vitro from 2 cells to expanded blastocyst. The blastocyst is composed of an outer polarized epithelium, the trophectoderm, and an inner cell mass (ICM), a non-epithelial cell cluster which gives rise to the embryo proper. The trophectoderm which forms most of the placental lineages, protects the developing ICM, generates the blastocoel and mediates implantation to establish pregnancy.

Epithelial features are acquired progressively by blastomeres from the 8-cell stage when E-cadherin mediated compaction occurs (reviewed in refs. 44, 45) and cytokeratins (synthesized approximately at the 4-cell stage) assemble into filaments.[46,47] Focal tight junction adhesion begins subsequent to E-cadherin adhesion in the 8-cell embryo and increases in molecular complexity during the 8-cell and 16-cell cycles to form a continuous seal in the outer cells at the 32-cell stage.[48-50] Blastocoel formation occurs initially by the intercellular accumulation of water and continues because water moves across the trophectoderm into the expanding blastocoel along an osmotic gradient of Na^+ ions established by vectorial transport.[51,52]

Desmosomes can be observed at the lateral cell membranes of trophectoderm cells at the 32- to 64-cell stage[53-55] at the onset of blastocoel formation at approximately

90-96 hours post-hCG.[12] Structural integrity in the trophectoderm is clearly important for its functional capacity as the blastocoel expands. Desmosomes are believed to be important in tissue cohesion and stability (see chapter 7) so it has been hypothesized that one of their functions is to strengthen the trophectoderm[12] and thus prevent blastocyst expansion becoming blastocyst explosion! In addition, it is possible that desmosomal adhesion provides a signal to trophectoderm cells which has functional consequences for cavitation or further differentiation of trophectoderm cells.

The sequential synthesis and membrane localization of the more abundant protein constituents of desmosomes have been studied using a combination of fluorescent antibody staining and immunoprecipitation of metabolically labeled proteins.[12] Synthesis of plakoglobin is detected at the 8-cell stage at the time of compaction (68-72 hours post-hCG), desmoplakins at the morula stage (80-84 hours post-hCG) with desmocollins and desmoglein at the late morula stage (90-96 hours post-hCG), just prior to the appearance of trophectoderm specific, punctate membrane staining and desmosome assembly. These observations reveal a temporal expression program which is underway from compaction in the 8-cell embryo and results in desmosome assembly in the trophectoderm of the nascent blastocyst.

During this period of differentiation the later translation of desmocollins and desmoglein appears to facilitate cell-type specific desmosome assembly.[12] The glycoproteins presumably provide a membrane attachment site for cytokeratins and are involved in direct or indirect interactions with the previously synthesized components desmoplakins and plakoglobin resulting in junction assembly. The ultrastructural localization of desmoplakin suggests that one of its functions is to attach intermediate filaments to desmosome membrane-associated proteins.[56] Detailed studies of the structure and molecular interactions of desmoplakin 1 indicate that it mediates specific associations with intermediate filaments at the desmosomal cell membrane domain,[57-62] possibly interacting with desmocollin (see chapter 3). As discussed earlier, chimeric proteins comprising of the cytoplasmic tails of DSC 'a' form spliced to the transmembrane domain of connexin 32 gap junction protein were able to promote assembly of dense submembraneous plaques in A431 cells containing desmoplakin and plakoglobin in association with intermediate filaments.[41] Desmoplakin staining in the preimplantation embryo was detected at the cell membrane coincident with membrane localization of the desmosomal cadherins even though its synthesis was detected earlier by metabolic labeling. Whether desmoplakin localizes to the membrane before or after it becomes associated with intermediate filaments is not clear. There are discrepancies in the literature relating to this point.[5,30,63-66] However, the embryo data are consistent with the idea that it is a requirement for desmoplakin to bind directly or indirectly to one of the desmosomal cadherins when it localizes at the membrane.

The pattern of plakoglobin staining in the late morula differed from the other desmosomal components in that it localized at cell membranes earlier, in a linear pattern and in the ICM in addition to the trophectoderm. In later blastocysts staining became punctate and confined to the trophectoderm.[12] Plakoglobin has been localized in adherens junctions and desmosomes[67-69] indicating a broader role in the coordination of cell-cell adhesion. Coprecipitation and blot overlay experiments have shown that plakoglobin interacts with the cytoplasmic domains of the desmogleins, DSG1 and DSG3[72,73] (see chapter 4) and E-cadherin.[74] Initial linear membrane staining in the ICM and trophectoderm is consistent with plakoglobin binding to E-cadherin since its distribution resembles E-cadherin in morulae and early blastocysts.[70,71]

The maturation of plakoglobin staining to a punctate trophectoderm-specific pattern presumably represents desmosome-associated plakoglobin, most likely plakoglobin which is bound to desmoglein. Co-immunoprecipitation and laser confocal microscopy studies are required to confirm these interpretations. In the mouse preimplantation embryo desmoglein isoform expression is not established at present.

Recently Hinck et al[76] have shown that in MDCK cells plakoglobin binds to E-cadherin soon after synthesis prior to localization at the membrane and incorporation of α-catenin into the complex. β-catenin also binds to newly synthesized E-cadherin[76,77] but is bound to a mutually exclusive population of E-cadherin molecules.[76] Plakoglobin appears to be excluded from mature adherens junctions in 8 day cultures of MDCK cells and Näthke et al[78] suggest that plakoglobin may dissociate from E-cadherin, exchanging with β-catenin or that E-cadherin/plakoglobin complexes may be degraded. This concurs with the observation of Fleming et al[12] that in later blastocysts plakoglobin was unlikely to define adherens junctions in the ICM since they are punctate and persist throughout blastocyst expansion whilst plakoglobin sites are linear and transitory.

It is becoming clear that the catenins including plakoglobin are involved in processes regulating cell adhesion, differentiation and proliferation[83,96] (see chapter 4). The E-cadherin/catenin system is important in localizing the Na^+, K^+-ATPase to basolateral membranes of polarized epithelial cells including mural trophectoderm[52] and MDCK cells.[79] Catenins have been shown to interact with other proteins including the APC gene product, a protein that is concerned with growth regulation in colorectal epithelia.[80,81] Näthke et al[78] have proposed a model based on careful cross-linking, immunoprecipitation and extraction studies whereby equilibria between 'free' and bound pools of catenins including plakoglobin, interacting with cadherins and other proteins may regulate cellular responses to extracellular signals for adhesion or proliferation. It is intriguing that this group have reported that plakoglobin may bind to the epidermal growth factor receptor (EGF-R) in the lateral membranes of MDCK cells.[76] EGF-R localizes to the trophectoderm and ICM in a similar pattern to plakoglobin so the early linear staining of plakoglobin could be EGF-R associated. Antisense to EGF-R has been shown to reduce the cavitation rate in embryos[84] (J.E. Collins and T.P. Fleming, unpublished observations). Although speculative it appears likely that plakoglobin is part of a signaling network involved in coordinating the differentiation of the trophectoderm and/or cavitation.

The synthesis of desmosomal cadherins in preimplantation embryos was detected just prior to desmosome assembly. Therefore, one potential control mechanism regulating this event could be the availability of these proteins. Thus translation or transcription of DSG or DSC glycoproteins could be used to regulate the timing of desmosome formation. Transcription studies using reverse transcriptase mediated polymerase chain reaction with poly A+ RNA from unfertilized eggs and mid-blastocysts show that two alternatively spliced mRNAs, (a) and (b), of the DSC2 gene are produced (J.E. Collins, J. Lorimer, D. Garrod, S. Pidsley, R. Buxton and T. Fleming, unpublished data). The cDNA sequence of mouse DSC2 has been determined from clones[40,85] from an 8.5 day gestational embryo library.[87] Mouse DSC3, which has been partially sequenced (J. Lorimer and D. Garrod, unpublished data) and would be recognized by the primers used to detect DSC2 transcripts was not found in these samples.

Transcripts were detected in unfertilized eggs and all stages up to precompact 8-cells but compact 8-cell embryos were always negative (J.E. Collins, J. Lorimer, D. Garrod, S. Pidsley, R. Buxton and T. Fleming, unpublished data). DSC2 transcripts were detected again in some 16-cell embryos and in all blastocysts. Single cell analysis showed the presence of transcripts in all 4-cell embryos (n = 5), 41% (n = 17) precompact 8-cell embryos, 0% (n = 15) compact 8-cell embryos, 46% (n = 13) 16-cell embryos and all (n = 45) early blastocysts. Expression of transcripts from the embryonic genome during 16- and 32-cell stages occurs immediately before detection of desmocollin protein by immunoprecipitation.[12] These data indicate that assembly of the earliest desmosomes in the embryo is conditional on transcription and subsequent translation of DSC2 and therefore transcription timing provides a component of the regulatory mechanism. Transcripts detected in unfertilized eggs are maternal in origin and untranslated since no evidence of desmocollin translation was found at this time.[12] Maternal DSC2 mRNA appears to be inherited and remain under translational arrest whilst being degraded by the early cleaving embryo. Most maternal mRNA is degraded up to and during the 2-cell stage[88,89] but degradation rates[90] and detection sensitivity varies; Rappolee et al[91] could detect maternal TGF-α transcripts until the 8-cell stage using polymerase chain reaction methods consistent with our data.

What are the signals regulating transcription of DSC2 in the early embryo? Studies on cultured epithelial cells have shown that adhesion of E-cadherin is a prerequisite for normal development of tight and desmosomal junctions[6] so it was pertinent to examine whether inhibition of E-cadherin mediated adhesion would alter the timing of DSC2 transcription. Delaying compaction by incubation of early 8-cell embryos in ECCD-1[92] or complete inhibition using Ca^{2+}-free medium had no effect on the timing of DSC2 mRNA expression. In addition, DSC2 protein expression was unaffected. However, ECCD-1 produced disorganization of DSC2 and desmoplakin in these blastocyst equivalent stage embryos and in the absence of Ca^{2+} no evidence of desmosome assembly was observed at cell membrane sites (J.E. Collins, J. Lorimer, D. Garrod, S. Pidsley, R. Buxton and T. Fleming, unpublished data). These results indicate that compaction mediated by E-cadherin is important for desmosomal organization but is not a requirement for activation of transcription of the DSC2 gene in the preimplantation embryo, although it is possible that E-cadherin can function in ways which do not depend on it being in an adhesive conformation.

Recently, our studies have indicated that DSC2 transcription is initiated in cells which possess a free cell surface (J.E. Collins, D.R. Garrod and T.P. Fleming, unpublished data). In the preimplantation embryo these are the outer cells which form the trophectoderm and in later blastocysts, the presumptive primitive endoderm which is exposed to the blastocoel cavity after retraction of trophectodermal processes.[93] The presence or loss of contact-free cell surfaces has also been implicated in mechanisms regulating differential translation of the proteins ZO-1 and cingulin in the early embryo.[50,94] The molecular basis of these cellular interactions is currently under study in our laboratory.

MODULATION OF DESMOSOMAL ADHESION IN EPITHELIAL CELLS

We have recently found some evidence that desmosomes are rapidly responsive to signals dependent upon intercellular contact and protein kinase C. Thus, our previous work showed that desmosomal adhesion in keratinocytes and MDCK cells was initially calcium-dependent but matured to become calcium-independent.[5,29,43] This is surprising since the desmosomal adhesion molecules are members of the cadherin superfamily of calcium-dependent molecules. We have shown recently that this maturational change is entirely dependent upon confluency between the cells and is rapidly reversed to calcium-dependence if a cultured monolayer is wounded to create a free surface (S. Lloyd, I. Wise, T. Fleming and D. Garrod, unpublished observations). Modulation between calcium dependence and independence is also rapidly affected by activation of protein kinase C (which promotes calcium-dependence) and PKC inhibition (which promotes calcium-independence). These observations have not yet been extended to cells in vivo, but the possibility exists that where free edges are formed in tissue (e.g. wounding, tumor invasion, embryonic development) cells may modulate the adhesive properties of their desmosomes through activation of signaling pathways.

CONCLUSION

The detailed characterization of expression of desmocollins and the availability of isoform specific antibodies and cDNAs facilitates functional studies at a molecular level. We need to confirm directly that the desmocollins are adhesive glycoproteins. We wish to elucidate the proteins with which they interact both extracellularly and cytoplasmically. For example do they adhere homophilically or is the mechanism a heterophilic desmocollin/desmoglein interaction? What is the significance of differential isoform expression? Can the desmosome transduce information to the cell interior regarding its adhesive state? Ultimately, we wish to understand how adhesion is regulated. Creating free edges in cell monolayers or providing enclosed cells in the early embryo with free cell surfaces affects regulation of desmosomal adhesion. Studying the mechanisms involved in these processes will provide fascinating insights into the biology of desmosomal adhesion.

ACKNOWLEDGMENTS

We wish to thank Kevin Yue, Jo Lorimer, Tom Fleming, Sue Lloyd, Irene Wise and Kevin Legan for making unpublished data available for this manuscript. In addition, we thank Mark Hay and Jeanne Rodgers for technical assistance. The work was supported by grants from the Cancer Research Campaign, the Medical Research Council and the Wellcome Trust.

REFERENCES

1. Skerrow CJ, Matoltsy AG. Isolation of epidermal desmosomes. J Cell Biol 1974a; 63:515-523.
2. Gorbsky G, Steinberg, MS. Isolation of the intercellular glycoproteins of desmosomes. J Cell Biol 1981; 90:242-248.
3. Koch PJ, Walsh MJ, Schmelz M, et al. Identification of desmoglein, a constitutive desmosomal glycoprotein, as a member of the cadherin family of cell adhesion molecules. Eur J Cell Biol 1990; 53:1-12.
4. Holton JL, Kenny TP, Legan PK, et al. Desmosomal glycoproteins 2 and 3 (desmocollins) show N-terminal similarity to calcium-dependent cell-cell adhesion molecules. J Cell Sci 1990; 97:239-246.
5. Mattey DL, Garrod DR. Calcium-induced desmosome formation in cultured kidney epithelial cells. J Cell Sci 1986; 85:95-111.
6. Gumbiner B, Stevenson B, Grimaldi A. The role of the cell adhesion molecule uvomorulin in the formation and maintenance of the epithelial junctional complex. J Cell Biol 1988; 107:1575-1587.
7. Buxton RS, Cowin P, Franke WW, et al. Nomenclature of the desmosomal cadherins. J Cell Biol 1993; 121:481-483.
8. Arnemann J, Sullivan KH, Magee AI, et al. Stratification related expression of isoforms of the desmosomal cadherins in human epidermis. J Cell Sci 1993; 104:741-750.
9. Theis DG, Koch PJ, Franke WW. Differential synthesis of type 1 and type 2 desmocollin mRNAs in human stratified epithelia. Int J Dev Biol 1993; 37:101-110.
10. Legan PK, Yue KKM, Chidgey MAJ, et al. The bovine desmocollin family: A new gene and expression patterns reflecting epithelial cell proliferation and differentiation. J Cell Biol 1994; 126:507-518.
11. Garrod DR, Parrish EP, Mattey DL, et al. Desmosomes. In: Edelman GM, Cunningham, Thiery JP, eds. Morphoregulatory Molecules. Chichester: John Wiley & Sons, 1990:315-339.
12. Fleming TP, Garrod DR, Elsmore A. Desmosome Biogenesis in the Mouse Preimplantation Embryo. Development 1991; 112:527-539.
13. Yoshida M, Romberger DJ, Illig MG, et al. Transformation growth factor-β stimulates expression of desmosomal proteins in bronchial epithelial cells. Am J Respir Cell Mol Biol 1992; 6:439-445.
14. Christensen TG, Burke B, Dexter DL, et al. Ultrastructural evidence of dimethylformamide-induced differentiation of cultured human colon carcinoma cells. Cancer 1985; 56:1559-1565.
15. Garrod DR. The adhesions of epithelial cells. In: Marthy HJ. ed. NATO Advanced Stud-

ies Institute. Series A. Life Sci 1985; 99:43-83.
16. Garrod DR. Desmosomes and hemidesmosomes. Curr Opin Cell Biol 1993; 5:30-40.
17. Skerrow CJ, Matoltsy AG. Chemical characterisation of isolated epidermal desmosomes. J Cell Biol 1974b; 673:524-531.
18. Mueller H, Franke WW. Biochemical and immunological characterisation of desmoplakins I and II, the major polypeptides of the desmosomal plaque. J Mol Biol 1983; 163:647-671.
19. Cowin P, Garrod DR. Antibodies to epithelial desmosomes show wide tissue and species cross-reactivity. Nature (Lond.) 1983; 302:148-150.
20. Cohen SM, Gorbsky G, Steinberg MS. Immunochemical characterisation of related families of glycoproteins in desmosomes. J Biol Chem 1983; 258:2621-2627.
21. Cowin P, Mattey DL, Garrod DR. Identification of desmosomal surface components (desmocollins) and inhibition of desmosome formation by specific fab'. J Cell Sci 1984; 70:41-60.
22. Kapprell HP, Cowin P, Franke WW, et al. Biochemical characterisation of desmosomal proteins isolated from bovine muzzle epidermis: amino acid and carbohydrate composition. Eur J Cell Biol 1985; 36:217-219.
23. Cowin P, Mattey DL, Garrod DR. Distribution of desmosomal components in the tissues of vertebrates, studied by fluorescent antibody staining. J Cell Sci 1984; 66:119-132.
24. Suhrbier A, Garrod DR. An investigation of the molecular components of desmosomes in epithelial cells of vertebrates. J Cell Sci 1986; 81:223-242.
25. Parrish EP, Garrod DR, Mattey DL, et al. Mouse antisera specific for desmosomal adhesion molecules of suprabasal skin cells, meninges and meningioma. Proc Natl Acad Sci 1986; 83:2657-2661.
26. Mattey DL, Suhribier A, Parrish E, et al. Recognition: calcium and the control of desmosome formation. In: Bock G, Clark S. eds. Junctional complexes of epithelial cells (Ciba Found Symp 125) Chichester, John Wiley and Sons, 1987; 49-65.
27. Steinberg MS, Shida H, Giudice GJ, et al. On the molecular organisation, diversity and functions of desmosomal proteins. In: Bock G, Clark S, eds. Junctional complexes of epithelial cells (CIBA Found Symp 125) Chichester, John Wiley and Sons, 1987; 3-25.
28. Hennings H, Holbrook KA. Calcium regulation of cell-cell contact and differentiation of epidermal cells in culture. Exp Cell Res 1983; 143:127-142.
29. Watt FM, Mattey DL, Garrod DR. Calcium induced reorganisation of desmosomal components in cultured human keratinocytes. J Cell Sci 1984; 99:2221-2215.
30. Bologna M, Allen R, Dulbecco R. Organisation of cytokeratin bundles by desmosomes in rat mammary cells. J Cell Biol 1986; 102:560-567.
31. Parrish EP, Marston JE, Mattey DL, et al. Size heterogeneity, phosphorylation and transmembrane organisation of desmosomal glycoproteins 2 and 3 (desmocollins) in MDCK cells. J Cell Sci 1990; 96:239-248.
32. Koch PJ, Goldschmidt MD, Walsh MJ, et al. Amino acid sequence of bovine muzzle epithelial desmocollin derived from cloned cDNA; A novel subtype of desmosomal cadherins. Differentiation 1991; 47:29-36.
33. Collins JE, Legan PK, Kenny TP, et al. Cloning and sequence analysis of desmosomal glycoproteins 2 and 3 (desmocollins): cadherin-like desmosomal adhesion molecules with heterogeneous cytoplasmic domains. J Cell Biol 1991; 113:381-391.
34. Parker AE, Wheeler GN, Arnemann J, et al. Desmosomal glycoproteins II and III: cadherin-like junctional molecules generated by alternative splicing. J Biol Chem 1991; 266:10438-10445.
35. Mechanic S, Raynor K, Hill JE, et al. Desmocollins form a subset of the cadherin family of cell adhesion molecules. Proc Natl Acad Sci USA 1991; 88:4476-4480.
36. Koch PJ, Goldschmidt MJ, Walsh R, et al. Complete amino acid sequence of the epidermal desmoglein precursor polypeptide and identification of a second type of desmoglein gene. Eur J Cell Biol 1991; 55:200-208.

37. King IA, Magee AI, Rees DA et al. Keratinization is associated with the expression of a new protein related to the desmosomal cadherins DGII/III. FEBS Lett 1991; 286:9-12.
38. Koch PJ, Goldschmidt MD, Zimbelmann R, et al. Complexity and expression patterns of the desmosomal cadherins. Proc Natl Acad Sci USA 1992; 89:353-357.
39. Schäfer S, Koch PJ, Franke WW. Identification of the ubiquitous human desmoglein, DSG2, and the expresion catalogue of the desmoglein subfamily of desmosomal cadherins. Exp Cell Res 1994; 211:391-399.
40. Lorimer JE, Hall LH, Collins JE, et al. Cloning, sequence analysis and expression pattern of a mouse desmocollin: a cadherin-like adhesion molecule. Mol Membrane Biol in press.
41. Troyanovsky, SM, Eshkind LG, Troyanovsky RB, et al. Contributions of cytoplasmic domains of desmosomal cadherins to desmosome asembly and intermediate filament anchorage. Cell 1993; 72:561-574.
42. Lavker RM, Sun TT. Epidermal stem cells. J Invest Dermatol 1983; 81:121s-127s.
43. Mattey DL, Garrod DR. Splitting and internalisation of the desmosomes of cultured kidney epithelial cells by reduction in calcium concentration. J Cell Sci 1986; 85:113-124.
44. Fleming TP, Johnson MH. From egg to epithelium. Ann Rev Cell Biol 1988; 4:459-485.
45. Pratt HPM. Marking time and making space: chronology and topography in the early mouse embryo. Int Rev Cytol 1989; 117:99-130.
46. Oshima RG, Howe WE, Klier FG, et al. Intermediate filament protein synthesis in preimplantation murine embryos. Dev Biol 1983; 99:447-455.
47. Chisholm JC, Houliston E. Cytokeratin filament assembly in the preimplantation mouse embryo. Development 1987; 101:565-582.
48. Fleming TP, McConnell J, Johnson MH, et al. Development of tight junctions de novo in the mouse early embryo: control of assembly of the tight junction specific protein, ZO1. J Cell Biol 1989; 108:1407-1418.
49. Fleming TP, Hay MJ, Javed Q, et al. Localisation of tight junction protein cingulin is temporally and spatially regulated during early mouse development. Development 1993; 117:1135-1144.
50. Javed Q, Fleming TP, Hay MJ, et al. Tight junction protein cingulin is expressed by material and embryonic genomes during early mouse development. Development 1993; 117:1145-1151.
51. Watson AJ, Kidder GM. Immunofluorescence assessment of the timing of appearance and cellular distribution of Na/K-ATPase during mouse embryogenesis. Devl Biol 1988; 126:80-90.
52. Watson AJ, Damsky CH, Kidder GM. Differentiation of an epithelium: factors affecting the polarized distribution of Na+, Ka+-ATPase in mouse trophectoderm. Dev Biol 1990; 141:104-114.
53. Ducibella T, Albertini DF, Anderson E, et al. The preimplantation mammalian embryo: characterisation of intercellular junctions and their appearance during development. Dev Biol 1975; 45:231-250.
54. Magnuson T, Demsey A, Stackpole C. Characterization of intercellular junctions in the preimplantation mouse embryo by freeze-fracture and thin-section electron microscopy. Dev Biol 1975; 61:252-261.
55. Jackson BW, Grund C, Schmid E, et al. Formation of cytoskeletal elements during mouse embryogenesis. Differentiation 1980; 17:161-179.
56. Miller K, Mattey D, Measures H, et al. Localisation of the protein and glycoprotein components of bovine nasal epithelial desmosomes by immunoelectron microscopy. EMBO J 1987; 6:885-889.
57. O'Keefe EJ, Erickson HP, Bennett V. Desmoplakin I and desmoplakin II: purification and characterization. J Biol Chem 1989; 264:8310-8318.
58. Green KJ, Parry DAD, Steinert PM, et al. Structure of the human desmoplakins: implications for function in the desmosomal plaque. J Biol Chem 1990; 265:2603-2612.
59. Green KJ, Stappenbeck TS, Parry DAD, et al. Structure of desmoplakin and its association with intermediate filaments. J Der-

matol 1992; 19:765-769.

60. Virata MLA, Wagner RM, Parry DAD et al. Molecular structure of the human desmoplakin I and II amino terminus. Proc Natl Acad Sci 1992; 89:544-548.
61. Stappenbeck TS, Green KJ. The desmoplakin carboxyl terminus coaligns with and specifically disrupts intermediate filament networks when expressed in cultured cells. J Cell Biol 1992; 116:1197-1209.
62. Stappenbeck TS, Bornslaeger EA, Corcoran CM, et al. Functional analysis of desmoplakin domains: specification of the interaction with keratin versus vimentin intermediate filament networks. J Cell Biol 1993; 123:691-705.
63. Jones JCR, Goldman RD. Intermediate filaments and the initation of desmosome assembly. J Cell Biol 1985; 101:506-517.
64. Pasdar M, Nelson WJ. Kinetics of desmosome assembly in Madin-Darby canine kidney epithelial cells: temporal and spatial regulation of desmoplakin organization and stabilization upon cell-cell contact. I. Biochemical analysis. J Cell Biol 1988; 106:677-685.
65. Pasdar M, Nelson WJ. Kinetics of desmosome assembly in Madin- Darby canine kidney cells: temporal and spatial regulation of desmoplakin organization and stabilization upon cell-cell contact. II. Morphological analysis. J Cell Biol 1988; 106:687-695.
66. Trevor KT, Steben LS. Distribution of desmosomal proteins in F9 embryonal carcinoma cells and epithelial cell derivatives. J.Cell Sci 1992; 103:69-80.
67. Docherty RJ, Edwards JG, Garrod DR, et al. Chick embryonic pigmented retina is one of the group of epithelioid tissues that lack cytokeratins and desmosomes and have intermediate filaments composed of vimentin. J Cell Sci 1984; 71:61-74.
68. Cowin P, Kapprell H-P, Tamkun J. et al. Plakoglobin: a protein common to different kinds of intercellular adhering junctions. Cell 1986; 46:1063-1073.
69. Kapprell H-P, Cowin P, Franke WW. Biochemical characterization of the soluble form of the junctional plaque protein, plakoglobin, from different cell types. Eur J Biochem 1987; 166:505-517.
70. Vestweber D, Gossler A, Boller K, et al. Expression and distribution of cell adhesion molecule uvomorulin in mouse preimplantation embryos. Dev Biol 1987; 124:451-456.
71. Kemler R, Gossler A, Mansouri A, et al. The cell adhesion molecule uvomorulin. In: Edelman GM, Cunningham BA, Thiery JP, Eds. The Neurosciences Institute Publications Series. Chichester, John Wiley and Sons, 1990; pp 41-56.
72. Korman NJ, Eyre RW, Klaus-Kovtun V, et al. Demonstration of an adhering junction molecule (plakoglobin) in the autoantigens of pemphigus foliaceus and pemphigus vulgaris. N Engl J Med 1989; 321:631-635.
73. Mathur M, Goodwin L, Cowin P. The interactions of a desmosomal cadherin, DSG1, with plakoglobin. Journal of Biochemistry 1994; 269:1-6.
74. Knudsen KA, Wheelock MJ. Plakoglobin, or an 83-kD homologue distinct from beta-catenin, interacts with E-cadherin and N-cadherin. 1992; 118:671-679.
75. Vestweber D, Gossler A, Boller K, et al. Expression and distribution of cell adhesion molecule uvomorulin in mouse preimplantation embryos. Dev Biol 1987; 124:451-456.
76. Hinck L, Näthke IS, Papkoff J, et al. Dynamics of cadherin/catenin complex formation: Novel protein interactions and pathways of complex assembly. J Cell Biol 1994; 125:1327-1340.
77. Ozawa M, Kemler R. Molecular organization of the uvomorulin catenin complex. J Cell Biol 1992; 116:989-996.
78. Näthke IS, Hinck L, Swedlow JR, et al. Defining interactions and distributions of cadherin and catenin complexes in polarized epithelial cells. J Cell Biol 1994; 125:1341-1352.
79. Marrs JA, Napolitano EW, Murphy-Erdosh C, et al. Distinguishing roles of the membrane-cytoskeleton and cadherin mediated cell-cell adhesion in generating different Na^+,K^+-ATPase distributions in polarized epithelia. J Cell Biol 1993; 123:149-164.
80. Su L-K, Vogelstein B, Kinzler KW. Asso-

ciation of the APC tumor suppressor protein with catenins. Science (Wash. DC) 1993; 262:1734-1737.

81. Rubinfeld B, Souza B, Albert I, et al. Association of the APC gene product with β-catenin. Science (Wash. DC) 1993; 262:1731-1734.

82. Dardik A, Smith RM, Schultz RM. Colocalization of transforming growth factor-α and a functional epidermal growth factor receptor (EGFR) to the inner cell mass and preferential localization of the EGFR on the basolateral surface of the trophectoderm in the mouse blastocyst. Dev Biol 1992; 154:396-409.

83. Peifer M, Cancer, catenins, and cuticle pattern: a complex connection. Science 1993; 262:1667-1668.

84. Brice EC, Jie-Xin Wu, Muraro R, et al. Modulation of mouse preimplantation development by epidermal growth factor receptor antibodies, antisense RNA, and deoxyoligonucleotides. Developmental Genetics 1993; 14:174-184.

85. Buxton RS, Wheeler GN, Pidsley SC, et al. Mouse desmocollin (DSC3) and desmoglien (DSG1) genes are closely linked in the proximal region of chromosome 18. Genomics 1994; 21:510-516.

86. Lorimer JE, Hall LS, Clarke JP, et al. Cloning, sequence analysis and expression pattern of mouse desmocollin 2 (DSC 2): a cadherin-like adhesion molecule. Molecular Membrane Biol 1994; in press.

87. Fahrner K, Hogan BLM, Flavell RA. Transcription of H-2 and Qa genes in embryonic and adult mice. EMBO J. 1987; 6:1265-1271.

88. Giebelhaus DH, Heikkila JJ, Schultz GA. Changes in the quantity of histone and actin messenger RNA during the development of preimplantation mouse embryos. Dev Biol 1983; 98:148-154.

89. Bachvarova R, Cohen EM, De Leon V, et al. Amounts and modulation of actin mRNAs in mouse oocytes and embryos. Development 1989; 106:561-565.

90. Paynton BV, Rempel R, Bachvarova R. Changes in state of adenylation and time course of degradation of maternal mRNAs during oocyte maturation and early embryonic development in the mouse. Dev Biol 1988; 129:304-314.

91. Rappolee DA, Brenner CA, Schultz R, et al. Developmental expression of PDGF, TGF-α, and TGF-β genes in preimplantation mouse embryos. Science 1988; 241:1823-1825.

92. Yoshida-Noro C, Suzuki N, Takeichi M. Molecular nature of the calcium-dependent cell-cell adhesion system in mouse teratocarcinoma and embryonic cells studied with a monoclonal antibody. Dev Biol 1984; 101:19-27.

93. Fleming TP, Warren PD, Chisholm JC. Trophectodermal processes regulate the expression of totipotency within the inner cell mass of the mouse expending blastocyst. J Embryol Exp Morph 1984; 84:63-90.

94. Fleming TP, Hay MJ. Tissue-specific control of expression of the tight junction polypeptide ZO-1 in the mouse early embryo. Development 1991; 113:295-304.

95. King IA, Arnemann J, Spurr NK, et al. Cloning of the cDNA (DSC1) coding for human type 1 desmocollin and its assignment to chromosome 18. Genomics 1993; 18.185-189.

96. Peifer M, Berg S, Reynolds AB. A repeating amino acid motif shared by proteins with diverse cellular roles. Cell 1994; 76:789-791.

CHAPTER 3

Molecular Genetic Analysis of Desmosomal Proteins

Elayne A. Bornslaeger, Thaddeus S. Stappenbeck,
Andrew P. Kowalczyk, Helena L. Palka, Kathleen J. Green

INTRODUCTION

The integrity, morphology and ultimately the function of a tissue requires the incorporation of individual cells into an integrated unit. This is accomplished in part by specialized junctional complexes that function either in cell-substrate or cell-cell adhesion.[1] One type of intercellular adhesive junction is the desmosome. Desmosomes act not only in cell-cell adhesion but also provide cell surface attachment sites for the intermediate filament (IF) cytoskeleton. Because of these dual functions desmosomes play a key role in the establishment and maintenance of a transcellular IF network that is critical for the mechanical integrity of cell sheets and tissues. Most epithelial tissues prominently feature desmosomes in association with keratin IF. Desmosomes are also found in nonepithelial tissues such as the arachnoidal meninges and dendritic reticular cells of lymph nodes, where they are associated with vimentin IF and the intercalated discs of heart muscle where they are associated with desmin IF.[2]

The proper assembly and maintenance of the desmosome-IF complex is of great importance for the development and function of these tissues. For example, mutations in genes encoding IF proteins are linked to severe blistering diseases of the skin;[3-5] in addition, desmosomal constituents are the target antigens for several autoimmune blistering diseases (see chapter 7). Desmosome assembly and stability are modulated during such processes as wound healing[6-8] and metastasis.[9,10] We have observed that desmosome assembly or stability correlates inversely with the invasive potential of a series of tumor cell lines.[11] The picture of desmosomes that emerges is that they are highly resilient structures that resist mechanical stress yet they are also dynamic structures that respond to growth

Molecular Biology of Desmosomes and Hemidesmosomes, edited by Jane E. Collins and David R. Garrod.

factors and environmental cues.[12] To fully understand the assembly and maintenance of the desmosome-IF complex it will be important to understand the interactions between individual constituents of this complex and how these interactions are regulated.

MOLECULAR COMPOSITION OF DESMOSOMES

Desmosomes appear at the ultrastructural level (Fig. 3.1) to be composed of a membrane core region sandwiched between symmetrical electron dense cytoplasmic plaques.[2,13-19] Each tripartite plaque consists of an extremely electron dense outer plaque subjacent to the plasma membrane that is separated by an electron lucent zone from the fibrillar inner plaque through which IF bundles appear to loop. Several major desmosomal components have been identified. The membrane core of the desmosome contains the extracellular domains of the desmogleins and desmocollins, transmembrane glycoproteins that have recently been shown to be part of the cadherin family of cell-cell adhesion molecules.[20,21] By analogy with the classical cadherins the desmosomal cadherins are thought to mediate cell-cell adhesion, although this has not yet been demonstrated directly. The outer plaque is composed in large part of the cytoplasmic tails of the desmosomal cadherins as well as the 83 kd desmosomal cadherin associated protein plakoglobin (see chapter 4). The major components of the inner plaque are the most abundant plaque proteins of desmosomes, desmoplakins I and II (DPI and DPII).[22,23] Several other desmosome specific plaque components are present in lesser amounts or with a restricted tissue distribution. These include a 240 kd calmodulin binding protein that is expressed only in stratified epithelia; the human homologue is known as keratocalmin[24] and the bovine form as desmocalmin.[25] Another stratification specific desmosomal component is the 75 kd Band 6 protein, which is also known as DPIV.[26] A 140 kd plaque protein that is antigenically related to lamin-B has been described by Cartaud and colleagues.[27] (see Note Added in Proof section on page 49) A 140 kd protein described by Ouyang and Sugrue[28] has been localized to the innermost region of the plaque, even deeper than DP. The possibility that these 140 kd proteins are identical has not been ruled out. Additional

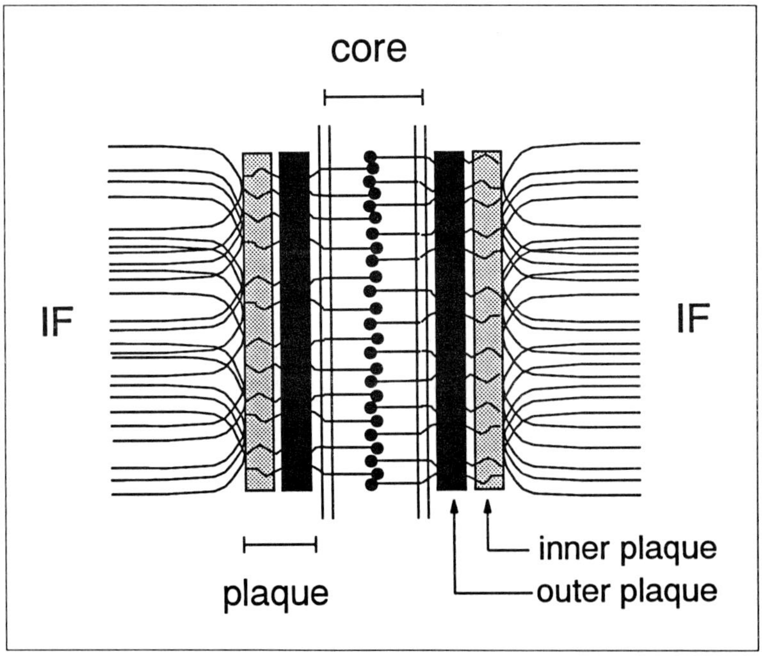

Fig. 3.1. Schematic representation of desmosome ultrastructure. Desmosomes appear at the ultrastructural level to be composed of a membrane core region sandwiched between symmetrical electron dense cytoplasmic plaques. Each tripartite plaque consists of an extremely electron dense outer plaque subjacent to the plasma membrane that is separated by a relatively electron lucent zone from the fibrillar inner plaque through which IF bundles appear to loop.

desmosomal components include the IF associated proteins (IFAPs) plectin[29] and IFAP 300.[30] As will be described below, although these proteins localize to desmosomes they also exhibit a more general cellular distribution.

Protein interactions between components of macromolecular complexes have often been established by traditional biochemical methodology, as is the case for the erythroid membrane cytoskeleton, for example.[1] Biochemical studies of the protein-protein interactions between desmosomal components have been limited to some extent by the insolubility of desmosomal components once they are assembled into desmosomes. Extraction and purification of desmosomal proteins requires denaturing conditions that may impair the ability of these proteins to interact in vitro as they would in vivo. Therefore we have chosen to use a molecular genetic approach to study the interactions between components of the desmosome-IF network.

DELINEATION OF DESMOPLAKIN FUNCTIONAL DOMAINS

The location of the DPs at the interface between the cellular IF network and the cytoplasmic tails of the desmosomal cadherins[31] suggests that DP may play a role in linking the IF cytoskeleton to the outer desmosomal plaque. Until recently little was known about DP structure or function.

Predicted Structure of Desmoplakin and Its Relatives

To obtain clues as to the function of DP and to acquire tools for the functional studies described below, we isolated and characterized DP cDNA and genomic clones. DPI and DPII are encoded by a single gene that is predicted to give rise to polypeptides of 332 and 260 kd, respectively.[32-36] DPI is a constitutive desmosomal component whereas DPII has a more restricted tissue distribution.[37] Our analysis of the predicted amino acid sequence of DP in conjunction with gel filtration, crosslinking and rotary shadowing studies of purified DP suggest that DPI is a homodimer consisting of a central α-helical coiled-coil rod domain of about 130 nm in length flanked by globular end domains.[32-34,38,39] Although DPII has globular end domains identical to those of DPI, due to alternative splicing of the DP transcript using an internal donor the DPII rod domain is only about 43 nm in length. DPII is likely to form dimers in vivo based on its predicted structure although in vitro evidence for this has not yet been reported as is the case for DPI.[39] The pattern of acidic and basic residues along the heptad repeat containing rod domain of DPI indicates that in addition to dimerizing it may also aggregate with itself or similar molecules into higher order filamentous structures. DP may therefore be a component of the fine 4-5 nm "traversing filaments" that appear to link IF with the plaque.[40]

The C-terminal globular domain of DP can be divided into three 176 amino acid subdomains, A, B and C, followed by a 68 residue "tail" at the very C-terminus of the molecule. Each subdomain is made up of a 38 residue repeating motif. Interestingly, this motif exhibits the same periodicity of acidic and basic residues as the 1B rod domain of IF proteins, raising the possibility that DP interacts with IF or IF-like proteins by virtue of their common periodicity of charged residues.[32,38] The N-terminal globular domain of DP is made up of a series of heptad repeats that are shorter and have a higher ratio of apolar to charged residues than the heptad stretches in the rod domain. Because of this, the N-terminal heptad repeats are predicted to form compact bundles rather than a coiled-coil rod.[34]

This basic structural arrangement of a central coiled-coil rod domain with globular end domains is found not just in DP but also in two other proteins that share regions of sequence homology with DP. One of these, the 230 kD bullous pemphigoid antigen (BP 230)[41-43] is a hemidesmosomal plaque protein and an antigen for autoantibodies in patients with the blistering skin disease, bullous pemphigoid (see chapter 7). The third family member is the intermediate filament associated protein (IFAP) plectin[44] which colocalizes with desmosomes although it also exhibits a more general cellular distribution. Based on partial sequence information and its immunological similarity to plectin, another possible member of this family is IFAP 300.[30] IFAP 300 colocalizes with both desmosomes and hemidesmosomes in epithelial cells as well as with vimentin IF in fibroblasts. The common structural characteristics and protein sequence homology among the three confirmed family members, DP, BPA and plectin, suggests that they function in a conserved manner in their interactions with IF networks. Regions of sequence divergence among family members must therefore be important for conferring functional specificity. For example, BP230 is specifically localized to the hemidesmosomal plaque and therefore probably contains hemidesmosome specific targeting sequences. Other than its localization in the recently reported "syndesmos" which are found only in the reticulothelium of lymph nodes,[45] DP is specifically localized in the desmosomal plaque and therefore is likely to contain desmosome specific targeting sequences.

Transfection Studies: Function of the DP C-terminus

DP is a candidate to link the IF network and the desmosomal outer plaque based on its location, its constitutive presence as a desmosomal component, and its sequence and structural similarity to known IFAPs. However, in vitro biochemical approaches have so far failed to identify a direct interaction between DP and IF or DP and proteins in the outer plaque such as desmosomal cadherins or plakoglobin.[39,46] In some cases this may be due to the use of harsh denaturing conditions to extract and purify desmosomal components. It is also possible that DP interactions with IF or the outer core proteins are not direct but are mediated by adapter proteins; alternatively, the interactions may be stabilized by accessory proteins. Such adapter or accessory proteins may be as yet unidentified desmosomal components. For these reasons we chose to use a molecular genetic approach to test potential interactions between DP and the desmosome-IF network. We transiently expressed DP domains or full length DP in cultured epithelial cells or fibroblasts and assessed the association of DP polypeptides with the cytoskeleton and the cell surface using immunocytochemical and biochemical approaches.[35,47] In some cases a peptide tag consisting of a 10 amino acid epitope of c-myc was introduced at the C-terminus so that ectopically expressed DP polypeptides and endogenous DP could be distinguished.[48]

Acidic and basic keratin pairs are thought to interact in part via the charge distribution along their 1B rod domains. Since the C-terminal subdomains of DP exhibit a similar distribution of acidic and basic residues we predicted that the C-terminal domain of DP may be able to interact with IF proteins. We therefore tested the ability of the DP C-terminal domain to interact with IF. DP polypeptides containing the DP C-terminal domain alone (DP.CT) or in conjunction with the rod domain (DP.ΔN) align specifically along both keratin and vimentin IF networks when expressed in cultured epithelial cells (see Fig. 3.2 for an example in HeLa cells). This alignment requires the presence of the DP C-terminus; in its absence, the rod domain does not interact with IF even though it is expressed at very high levels.[47] When expressed at high levels DP polypeptides containing the DP C-terminus disrupt both keratin and vimentin networks[35,47] (see Figs. 3.5A and B). The disruption is specific for IF; there are no apparent effects on either microfilaments or microtubules. These results indicate that the function of the DP C-terminus may be to interact with the cellular IF network, consistent with the idea that DP may play a role in linking the IF cytoskeleton to the desmosomal plaque.

It is unclear whether the interaction between DP and the cellular IF network

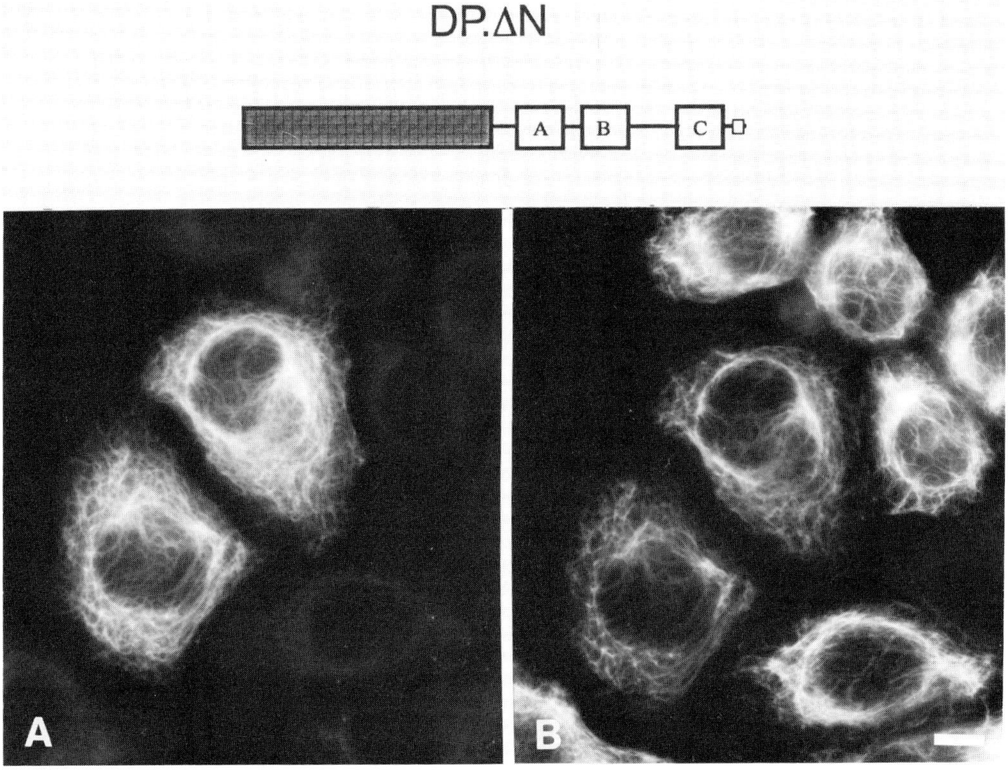

Fig. 3.2. Alignment of DP.ΔN with keratin IF in HeLa cells. Cells were transfected with a CMV expression vector encoding the central rod plus C-terminal globular domains of DP (DP.ΔN). Cells were methanol fixed and reacted with (A) a rabbit polyclonal against DP, NW38 or (B) a mouse monoclonal against keratin 18, KSB17.2. Note coalignment of the DP polypeptide with keratin IF in transfected cells. Although only the keratin pattern is shown, alignment along vimentin IF is also observed. Similar results are obtained for a construct encoding the DP C-terminal globular domain (DP.CT). Modified from Stappenbeck et al., J Cell Biol 1993; 123:691-705. Bar = 10 μm.

represents a direct interaction between the DP C-terminus and IF polypeptides or whether this interaction requires accessory proteins. DP.ΔN aligns with vimentin IF in 3T3 fibroblasts which presumably do not express desmosomal proteins,[47] suggesting that this interaction may be direct. Nevertheless, proteins found in both fibroblasts and epithelial cells such as IFAP 300 or plectin could be acting as adapters between DP and IF. IFAP 300 has recently been localized to both desmosomes and hemidesmosomes by morphological and biochemical criteria and hence has been suggested to be a linker between IF and both types of junctions.[30] Plectin also colocalizes with desmosomes.[29] It is interesting to note that Wiche and colleagues recently used a transient expression approach to examine the behavior of the plectin rod and C-terminal globular domains in cultured COS and PtK2 cells.[49] Similar to our observations that DP.CT or DP.ΔN align with and disrupt IF, the plectin C-terminal domain or rod plus C-terminus align with and eventually disrupt IF, whereas the rod domain alone does not exert this effect. Ectopically expressed DP.ROD appears as numerous cytoplasmic aggregates whereas the plectin rod domain exhibits a nuclear localization. The basis for this difference remains to be elucidated.

To further define the sequences required for the interaction of the DP C-terminus and IF networks, we analyzed C-terminal truncations of DP.ΔN.[35] Truncated proteins were transiently expressed in several cell lines so that alignment and disruption of vimentin (3T3 fibroblasts), keratin (HaCat keratinocytes) or both IF networks within the same cell (HeLa cells) could be assessed. We found that within the 68 residue C-terminal tail of DP are sequences that regulate specifically the interaction of DP with keratin IF networks. These residues do not seem to influence the interaction of DP with the vimentin IF network, however. Deletion of the entire 68 amino acid tail (DP.ΔN.Δ68) decreases the ability of DP.ΔN to interact with keratin networks (Figs. 3.3A and B) but has no effect on the interaction with vimentin networks (Figs. 3.3C and D). Critical for the interaction of DP.ΔN with keratin IF networks are residues 48-68 from the end of DP as a DP protein lacking the last 48 residues

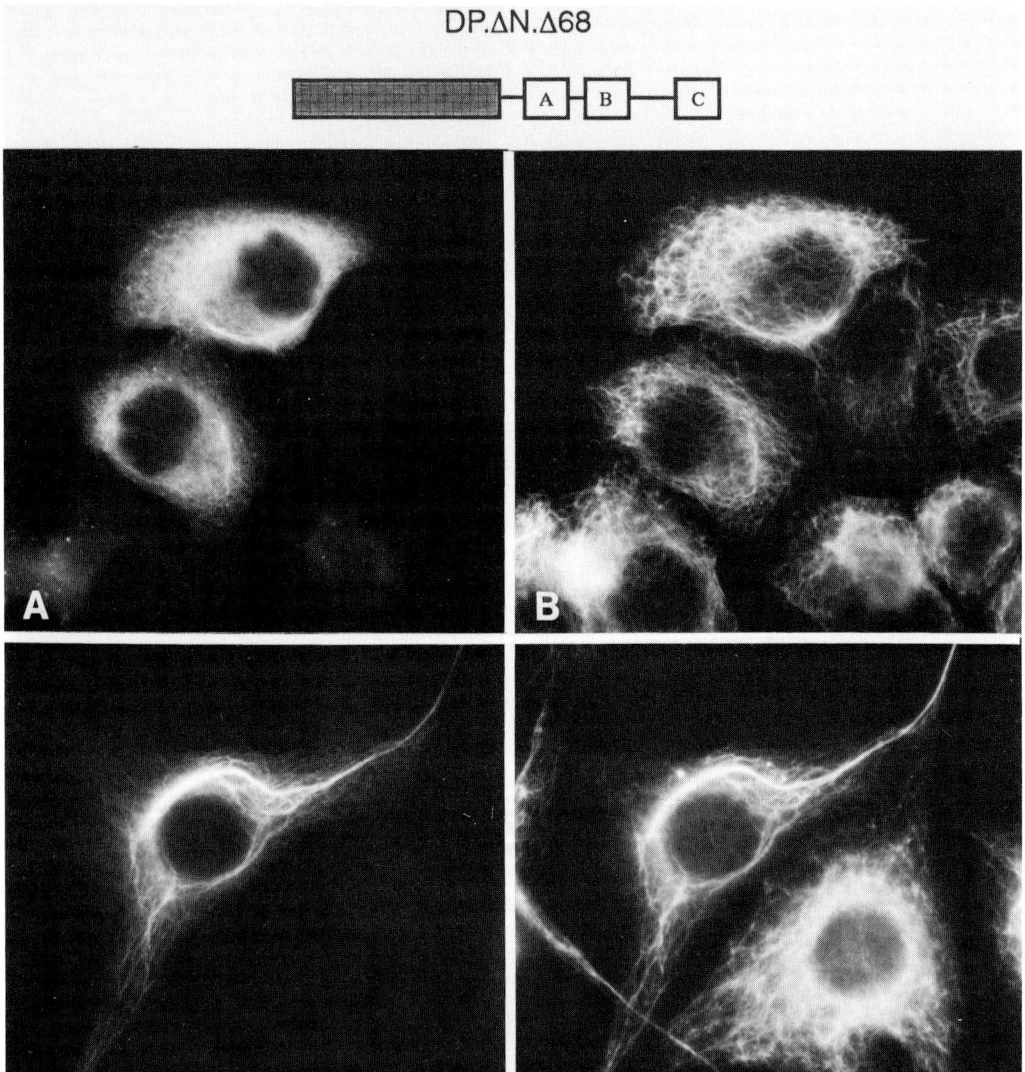

Fig. 3.3. Indirect double label immunofluorescence of HeLa cells or NIH-3T3 fibroblasts expressing DP.ΔN.Δ68. HeLa cells (A and B) or NIH-3T3 cells (C and D) were methanol fixed and reacted with a rabbit polyclonal against DP, NW38 (A and C), a mouse monoclonal against keratin 18, KSB17.2 (B) or a mouse monoclonal against vimentin, V9 (D). Note the lack of alignment of keratin IF with the DP polypeptide (A and B) whereas alignment with vimentin IF is observed (C and D). Modified from Stappenbeck et al., J Cell Biol 1993; 123:691-705. Bar = 10 μm.

(DP.ΔN.Δ48) does interact with keratin networks.

The presence or absence of the C-terminal 48-68 residues also affects the ability of DP.ΔN to be extracted from cells by nonionic detergents such as TritonX-100.[35] Normally very little DP.ΔN can be extracted from epithelial cells, presumably because it is associated with the keratin cytoskeletal system. Removal of the C-terminal 68 residues, however, results in greatly increased Triton solubility of the DP polypeptide. Furthermore, DP.ΔN is more Triton extractable in 3T3 fibroblasts, which contain only a Type III IF network, than in HeLa cells, which have both keratin and vimentin networks. Taken together, these observations indicate that resistance of DP.ΔN to Triton extraction correlates with its alignment along keratin IF networks. This suggests that DP may be more tightly associated with the keratin IF network than with the vimentin IF network and that the 68 C-terminal residues of DP are required for this tight association. The interaction of DP and keratin IF networks most likely requires not just these 68 residues but also two or more C-terminal subdomains; the 68 residue region alone or in conjunction with the C subdomain does not interact with keratin IF networks.

Although the C-terminal 48-68 residue region is not required for alignment of DP.ΔN with vimentin IF networks, at least two C-terminal subdomains are required for alignment, and efficient alignment only occurs with the addition of the rod domain.[35] The rod domain by itself does not align with or disrupt IF networks, but its addition to the C-terminus of DP presumably allows dimerization and effectively increases the number of C-terminal subdomains locally available for interaction with IF networks. The physiologic significance of these differences in the interaction of DP with keratin versus vimentin IF networks are unclear. Our results are consistent with previous observations that desmosomes may preferentially interact with keratin rather than vimentin IF in cells that have both IF systems.[50]

Regulation of DP Interaction with IF Networks

We have recently characterized an additional regulatory element within this 68 amino acid region.[72] Interestingly, truncation of the 28 most C-terminal residues of DP results in a protein (DP.ΔN.Δ28) that seems to align better with keratin networks than does DP.ΔN. A small amount of diffuse DP staining is commonly seen in cells expressing DP.ΔN whereas this is rarely observed for DP.ΔN.Δ28. In addition, DP.ΔN does not align well with keratin IF networks in cells treated with the adenylate cyclase activator forskolin to elevate intracellular cAMP levels (Fig. 3.4). Forskolin treatment has no effect on the alignment exhibited by DP.ΔN.Δ28, however. Within these 28 residues is a consensus site for cAMP-dependent protein kinase (protein kinase A). When the target serine residue within this protein kinase A site is altered to a glycine residue forskolin treatment has no effect on DP.ΔN alignment with keratin networks. These results suggest that protein kinase A-mediated phosphorylation of this target serine residue may decrease the affinity of the interaction between DP and keratin IF networks. Phosphopeptide mapping indicates that this DP C-terminal serine residue is indeed phosphorylated although it remains to be demonstrated that it is a specific target for protein kinase A or that its phosphorylation is solely responsible for the effects of forskolin on DP.ΔN alignment with keratin IF networks. These effects may also require phosphorylation of IF or DP-associated proteins. Nevertheless these results raise the intriguing possibility that the association of DP with IF networks may be regulated by the phosphorylation state of DP. This would further suggest that the linkage between the desmosomal plaque and the IF network may be influenced by DP phosphorylation, providing a potential means for regulating assembly of the desmosome-IF complex.

There are precedents for the idea that phosphorylation may regulate desmosome assembly. In addition to DP, other desmoso-

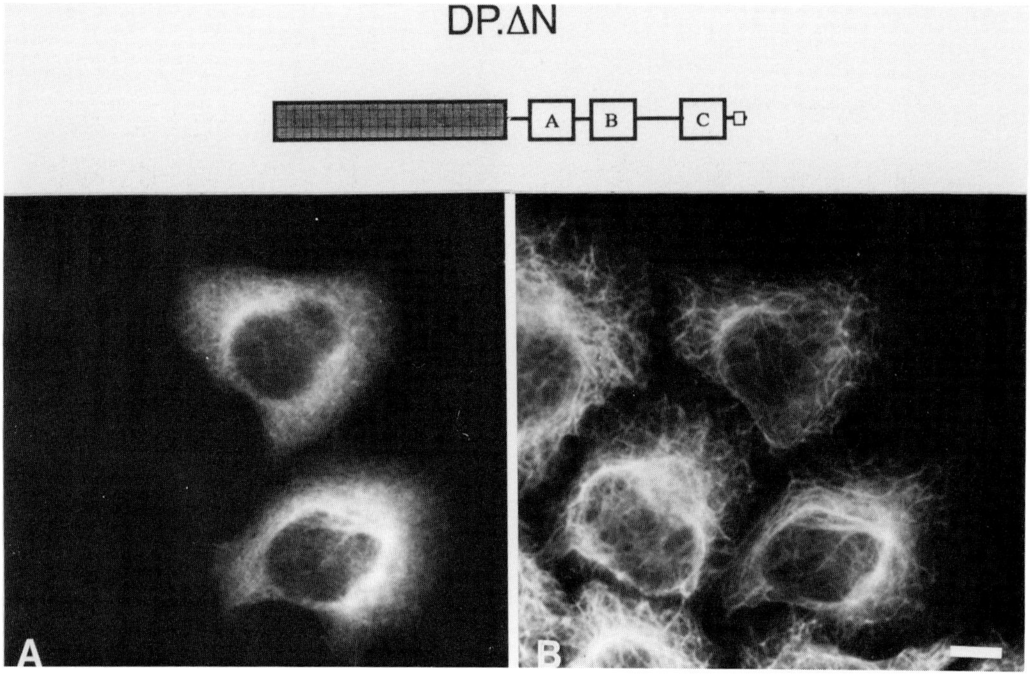

Fig. 3.4. Indirect double label immunofluorescence of HeLa cells expressing DP.ΔN in the presence of the adenylate cyclase activator forskolin. Cells were methanol fixed and reacted with (A) a rabbit polyclonal against DP, NW38 or (B) a mouse monoclonal against keratin 18, KSB17.2. Note the lack of coalignment of the DP polypeptide with keratin IF in transfected cells. Bar = 10 μm.

mal constituents such as desmocollin and plakoglobin are phosphoproteins.[50-53] Although protein kinase A has not been reported to regulate assembly, there are reports that modulation of protein kinase C activity can affect calcium induced desmosome assembly. Keratinocytes and other epithelial cells do not form desmosomes when they are cultured in low levels of extracellular calcium (< 0.1 mM); cell contact and desmosome assembly can be induced by switching cells into normal levels of extracellular calcium (1.8 mM). A four-fold increase in membrane bound protein kinase C activity has been observed within 15 minutes of such a calciumswitch in DJM-1 squamous carcinoma cells.[54] Furthermore, protein kinase C activation seems to promote desmosome assembly even in low calcium medium.[55] However, inhibitors of protein kinase C only partially inhibit desmosome formation following a switch from low to normal extracellular calcium.[55] An explanation for these apparently conflicting results remains to be elucidated as do the targets of protein kinase C activity in this system.

Function of the DP Rod

As mentioned above, the predicted structure of DPI indicates that it may contribute to the architecture of the plaque by aggregating with itself or similar molecules into higher order filamentous structures. By immunofluorescence, cells expressing high levels of either DP.CT or DP.ΔN exhibit large aggregates that contain both DP polypeptides and IF (Fig. 3.5A and B) as confirmed by immunogold electron microscopy (EM) (Figs. 3.5C and E).[47] In the presence or absence of the rod domain, however, the ultrastructure of these aggregates appears to be quite different. Aggregates containing DP.CT and IF are electron dense with no apparent filamentous substructure. Aggregates containing DP.ROD have a similar electron dense appearance, although

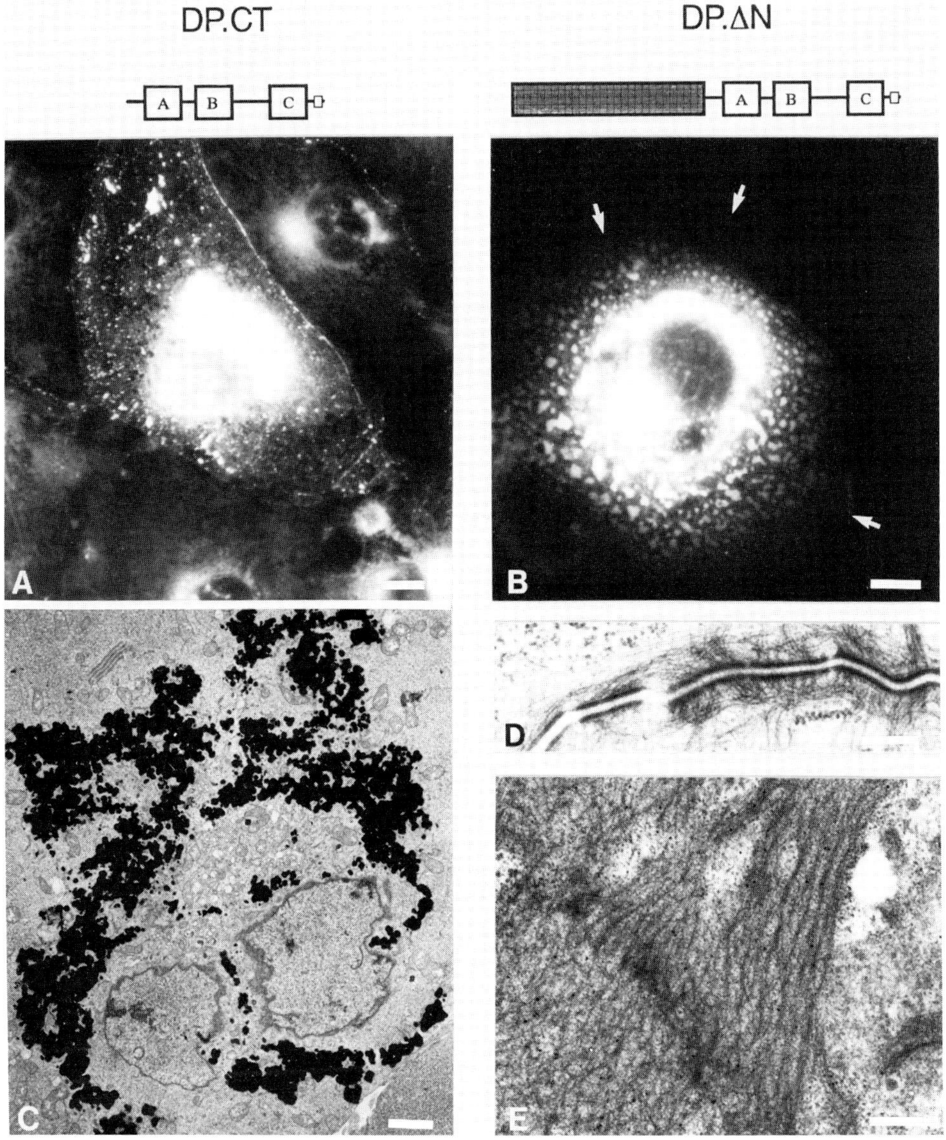

Fig. 3.5. Immunofluorescence and ultrastructural analysis of COS-7 cells expressing DP.CT.myc (A and C) or DP.ΔN.myc (B and E) and ultrastructural analysis of desmosomes formed between cultured primary mouse epidermal cells (D). Cells shown in A and B were reacted with either (A) a rabbit polyclonal against DP, NW38 or (B) a mouse monoclonal 9E10.2 directed against the c-myc epitope present at the C-terminus of the DP polypeptides. In similar cells the immunofluorescence patterns for the two antibodies are essentially identical with the exception that NW38 detects endogenous desmoplakin; note the desmosome staining at cell-cell borders in A. Arrows in B delineate the borders of the cell. Ultrastructural analysis was performed on cells fixed either for (C and D) conventional or (E) immunogold EM using 9E10.2. In C, clusters of very electron dense aggregates and no intact IF are present. See Stappenbeck and Green[47] for immunogold analysis of cells expressing DP.CT, demonstrating the presence of both DP and IF in these aggregates. In E, immunogold staining with 9E10.2 reveals the presence of c-myc tagged DP.ΔN in fibrous meshworks that have been shown in separate immunogold labeling experiments to contain vimentin IF as well.[47] These meshworks that contain DP.ΔN and IF resemble the desmosome plaques found in cultured cells such as the primary mouse epidermal cells shown in D. Modified from Stappenbeck and Green, J Cell Biol 1992; 116:1197-1209. For A and B, bar = 10 μm; for C, bar = 2.5 μm; for D and E, bar = 0.25 μm.

they do not contain IF. In the presence of DP proteins containing both the rod and C-terminal domains (DP.ΔN), however, aggregates that contain both ectopically expressed DP and IF appear as a fine filamentous meshwork. Importantly, this meshwork is only seen in aggregates that contain IF; DP.ΔN aggregates are sometimes seen that lack IF, and these have an amorphous nonfilamentous appearance. The filaments that make up this meshwork do not resemble the 8-10 nm IF found in neighboring nontransfected cells. Instead, they resemble the meshwork of 4-5 nm "fine filaments" or "traversing filaments"[40] that make up the innermost desmosomal plaque (Fig. 3.5D). These results suggest that the DP rod domain contributes to the architecture of the desmosomal plaque. It can do so, however, only in conjunction with IF and only if conjoined with the DP C-terminal domain. Furthermore, our results suggest that IF may attach to the plaque not as intact full caliber filaments but within an anastomosing network of intimately associated IF and DP.

Function of DP N-terminus

Since endogenous DP is localized to the desmosomal plaque, ectopically expressed DP would be predicted to colocalize with endogenous desmosomes. Such localization is not seen for either DP.CT or DP.ΔN suggesting that sequences in the DP N-terminus are required to target DP to the cytoplasmic face of the desmosome. Supporting this idea, ectopically expressed full length DP does colocalize with desmosomes (Fig. 3.6). Interestingly, this desmosomal colocalization is seen only in some cells within a population whereas in others only alignment along IF is seen.[35] We do not yet know what causes the function of the DP C-terminus or N-terminus to be manifest preferentially in any given cell. The cell cycle status or state of des-

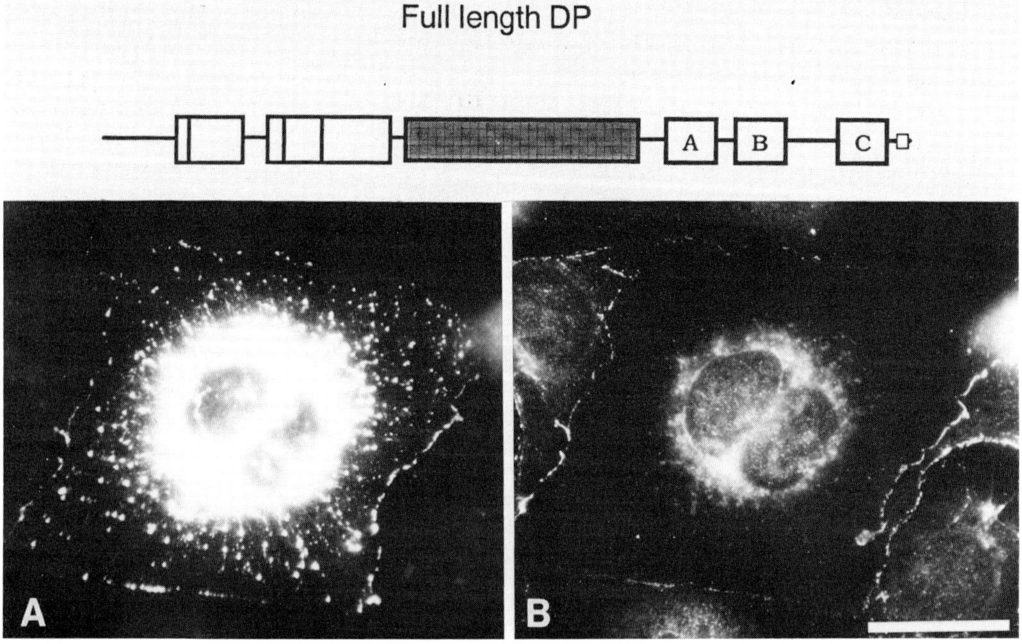

Fig. 3.6. Indirect double label immunofluorescence of COS-7 cells ectopically expressing full length c-myc tagged DPI. Cells were methanol fixed and reacted with the mouse monoclonal 9E10.2 directed against the c-myc tag (A) or a rabbit polyclonal against bovine desmoglein (B). Note the colocalization of 9E10.2 and desmoglein staining at cell-cell borders suggesting that tagged full length DP can associate with desmosomes. Modified from Stappenbeck et al., J Cell Biol 1993; 123:691-705. Bar = 10 μm.

mosome assembly within a cell may be important. Ectopically expressed DP may associate preferentially with newly forming desmosomes as found for example between daughter cells following cytokinesis.

The desmosome targeting function of the N-terminus requires the presence of the first 194 amino acids of DP; when this region is deleted only alignment with IF networks is seen.[35] This 194 amino acid region consists of sequences unique to DP that have not yet been reported in any other DP family members such as BPA or plectin,[38] consistent with the idea that this N-terminal region is required to target DP specifically to desmosomes. We do not yet know what sequences within this 194 residue region are critical for targeting DP to the plaque; the possibility that other regions of the DP N-terminus or rod domain also are involved remains to be tested. Nevertheless, the C-terminal domain of DP is apparently not required for cell surface localization of ectopically expressed DP. A truncated protein containing only the N-terminus and rod (DP.ΔC) is able to colocalize with desmosomes.[35] In fact, DP.ΔC is targeted to desmosomes in a higher percentage of cells than full length ectopically expressed DP. A possible explanation for this observation is that since DP.ΔC cannot interact with the IF network the function of the DP N-terminus would not have to compete with that of the C-terminus, as presumably occurs for ectopically expressed full length DP and endogenous DP as well.

Our analyses of the functional domains of DP are summarized in Figure 3.7. The presence of multiple functional domains within DP supports the idea that DP is a modular protein that links the IF cytoskeleton to the desmosomal outer plaque and also plays a role in organizing the inner plaque. The C-terminal globular domain of DP apparently directs the interaction of DP with the cytoplasmic IF network. However the last 68 residues of DP play a role only in its interaction with keratin and not vimentin networks. Residues 48-68 from the C-terminus are absolutely required for DP-keratin network interaction; furthermore the interaction of DP and the keratin network may be regulated by phosphorylation of a serine 23 residues from the DP C-terminus. The central coiled-coil rod domain of DP is implicated in generating the ultrastructural architecture of the desmosomal plaque, since the rod domain in conjunction with the DP C-terminus and in the presence of IF can be found in a meshwork that resembles the fibrous outer plaque of the desmosome. And finally, the N-terminus of DP is required to target DP to the cytoplasmic face of the desmosome. This targeting function does not require the C-terminus of DP since DP.ΔC appears to incorporate into desmosomes as well as, if not better than, ectopically expressed full length DP.

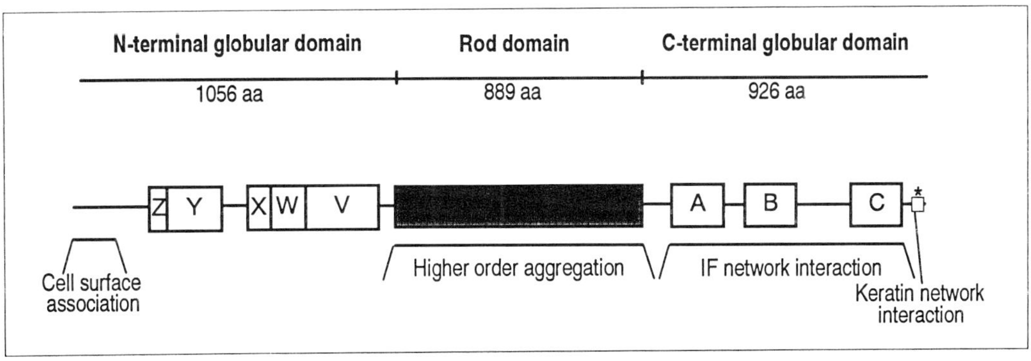

Fig. 3.7. Summary of domain analysis of DP. The three domains of DP are depicted as well as the number of amino acids in each. Below the line drawing are listed the domains of DP that are necessary for specific DP functions. The asterisk represents the serine residue present within the C-terminal 28 residues. For a description of the subdomains of the N-terminal globular domain (V-Z) see Green et al.[38]

DP ASSEMBLY INTO DESMOSOMES: ROLE OF INTERMEDIATE FILAMENTS?

Because DP.ΔC can apparently incorporate into desmosomes, our results suggest that an interaction between DP and IF is not required for DP to be targeted to the desmosome. We cannot rule out the possibility that IF play a regulatory role in the assembly of DP into the desmosomal plaque in vivo, however. There are conflicting reports in the literature concerning the role of IF in desmosome assembly. Some support the idea that DP requires an association with IF in order to be delivered to the cell surface and assembled into desmosomes. Jones and Goldman[56] examined the localization of DP and keratin IF during calcium induced desmosome assembly in primary mouse epidermal cells. When these cells were cultured in low levels of extracellular calcium immunofluorescence analysis revealed a colocalization of DP and keratin in a punctate juxtanuclear pattern. Following a switch to normal extracellular calcium levels and ensuing cell-cell contact, the DP spots were found to be relocated to the cell surface in parallel with extension of IF. Based on these observations, this group postulated that the movement of IF and associated desmosomal components to the cell surface is important for initiating desmosome assembly. Pasdar and Nelson[57,58] followed DP and its association with the IF cytoskeleton by both immunofluorescence and biochemical approaches during calcium induced desmosome assembly in MDCK cells. Their studies indicated that in low calcium conditions a detergent insoluble pool of DP appears as punctate spots that colocalize with keratin IF. This punctate staining pattern appeared to move rapidly to the cell periphery upon induction of cell-cell contact; these observations were likewise taken as support for the idea that IF may function as nucleation sites for the formation of DP-containing desmosome assembly intermediates in the cytoplasm. Trevor and Steben[59] examined the distribution of desmosomal proteins and keratin IF in retinoic acid-induced epithelial differentiation of F9 embryonal carcinoma cells. Although F9 cells normally lack desmosomes, in retinoic acid treated cultures about 50% of cells displayed punctate cell surface staining for DP whereas in other cells that lacked cell surface staining, spots of DP decorated keratin filaments. This group suggested the IF-associated DP spots they observed could represent DP desmosomal assembly intermediates prior to their delivery to the cell surface, which would be facilitated by their association with IF.

Several groups have proposed that desmosome assembly is initiated not along IF but instead at the cell surface. Bologna et al[60] followed the time course of keratin IF formation and localization of DP by immunofluorescence analysis of calcium induced desmosome assembly in a rat mammary epithelial line. They found that punctate DP staining at the cell surface, indicating desmosome assembly, apparently preceded attachment of cytokeratin bundles to the cell surface suggesting that desmosomes may serve to organize the IF cytoskeleton rather than the converse. Mattey and Garrod[61] have reported IF-associated spots containing DP in MDCK and MDBK cells cultured in low extracellular calcium. Analysis at the EM level suggested that these spots are desmosome derived vesicles that contain DP and other desmosomal components. Duden and Franke[62] performed a similar analysis of keratinocytes and several epithelial cell lines cultured in low calcium medium and also observed IF-associated vesicles containing DP that they interpreted as residual desmosome remnants that had been endocytosed following cell dissociation. More recently, Holm et al[63] and Burdett[64] have extended these studies by examining later events following desmosome internalization. Elucidation of the fate of desmosomal components following desmosome internalization and comparison with the routes taken by putative assembly intermediates should provide a much clearer picture of desmosome dynamics.

The apparent ability of DP.ΔC to incorporate into desmosomes in the absence of an interaction with the IF network is also consistent with reports that desmosomes can form normally in epithelial cells with impaired IF cytoskeletal systems. Vassar et al[65] reported normal desmosomes in transgenic mouse epidermis in which mutant keratin expression caused the disruption of keratin filaments and the formation of keratin aggregates. Baribault and Oshima[66] generated embryonic stem cells in which both alleles for keratin 8 had been inactivated. In these stem cells and their derivatives the type I keratins K18 and K19 were unable to pair with their type II partner K8 and therefore were unable to form keratin IF. When these stem cells were differentiated in culture they formed embryoid bodies with apparently normal epithelia. Immunofluorescence and ultrastructural analyses indicated that desmosomes were present in these epithelial cells even though IF were absent. Furthermore, mice homozygous for the inactivated K8 allele were able to develop at least to midgestation, with apparently normal epithelia.[67]

The ability of DP to interact with IF networks via its C-terminal domain and with the desmosomal plaque by virtue of its N-terminal domain suggests that these interactions must be regulated in vivo so that DP is localized preferentially to desmosomes and not along the cytoplasmic IF network. Our observation that in some cells ectopically expressed full length DP colocalizes with endogenous desmosomes whereas in other cells it aligns along IF networks suggests that mechanisms may exist to regulate the relative activity of the DP N- and C-terminal domains during desmosome assembly and maintenance. Our observations that forskolin treatment inhibits the association of DP.ΔN with keratin IF networks suggests that, at least for desmosomes that are associated with keratin IF networks, the phosphorylation state of the DP C-terminus may regulate the relative functional activity of the DP N- and C-termini. In this model, phosphorylation of the serine 23 residues from the DP C-terminus would inhibit DP from interacting with cytoplasmic IF networks and thus free it to localize to a "DP receptor" at the outer plaque via sequences in the DP N-terminus. Subsequent local dephosphorylation of plaque-associated DP would then allow the DP C-terminus to interact with the IF network at the cell surface.

ROLE OF DESMOSOMAL CADHERINS IN DESMOSOME ASSEMBLY

The N-terminus of DP allows it to localize to the desmosome. But what is the "receptor" for the DP N-terminus? Recent observations by Troyanovsky et al[68] (also see Note Added in Proof on page 49) support the idea that the cytoplasmic tails of the desmosomal cadherins, alone or in association with plakoglobin may be involved in recruiting DP to the desmosomal plaque. This group generated chimeric proteins containing the transmembrane domain of the gap junction protein connexin 32 and the cytoplasmic domains of desmosomal cadherins. When expressed in A431 epithelial cells a chimera containing the cytoplasmic domain of the longer spliced variant of desmocollin formed gap junction-like structures that recruited plakoglobin, DP and IF to the cell surface. Surprisingly, a connexin 32-desmoglein chimera behaved in a dominant negative fashion rather than recruiting plaque proteins and IF to the cell surface; there is no clear explanation for this observation. Nevertheless these results support the idea that the desmosomal cadherins play a critical role in initiating and regulating desmosome assembly.

Regulation of Plakoglobin Protein Expression by Desmosomal Cadherins

Understanding the regulation of desmosome assembly and stability ultimately depends on elucidating how interactions between specific desmosomal components are regulated. To this end we have begun to study the interactions between defined desmosomal constituents in the absence of other desmosomal proteins utilizing a system in which desmosomal components, alone or in combination, are stably expressed in mouse L-cell fibroblasts. Our initial studies have addressed interactions between the desmosomal cadherins and plakoglobin. Desmoglein and plakoglobin are known to coimmunoprecipitate from desmosome bearing cells under conditions that preserve protein-protein interactions[69] (see chapters 4 and 7), suggesting that they interact directly. Mathur and colleagues[70] have recently used blot overlay assays to demonstrate that plakoglobin and desmoglein interact directly and specifically and to map the sequences of desmoglein1 that are crucial for this interaction. Using the same approach, this group did not detect an interaction between desmocollin and plakoglobin, however.

We have generated mouse L-cell lines expressing the human desmosomal cadherins desmoglein or desmocollin alone or in combination with plakoglobin.[73] In cells coexpressing desmoglein and plakoglobin, the proteins colocalize to cell-cell interfaces (Fig. 3.8); the localization of plakoglobin to the cell surface presumably reflects an interaction between desmoglein and plakoglobin. We confirmed this by demonstrating that metabolically labeled desmoglein and plakoglobin coimmunoprecipitate from L-cell lysates under conditions that have been shown to maintain desmoglein-plakoglobin complexes. Evidence from L-cells coexpressing desmocollin and plakoglobin indicates for the first time that they interact in a similar manner.

Surprisingly, L-cells stably expressing plakoglobin but not a desmosomal cadherin express little or no plakoglobin protein even though the mRNA for human plakoglobin is present. When desmoglein is transiently expressed in such cells plakoglobin protein levels increase dramatically suggesting that desmoglein protein can regulate plakoglobin protein expression at a posttranscriptional level. A possible explanation for these results is that newly synthesized plakoglobin protein may be degraded rapidly unless it is bound to the cytoplasmic tail of desmoglein. Pulse-chase analysis to measure plakoglobin protein stability in the presence and absence of desmosomal cadherin supports this hypothesis. In lines coexpressing plakoglobin and desmoglein, or pakoglobin and desmocollin, plakoglobin protein is degraded much less rapidly than in L-cells expressing plakoglobin alone. These results raise the possibility that stabilization of plakoglobin via its interaction with desmosomal cadherins may be an important regulatory step in the sequence of

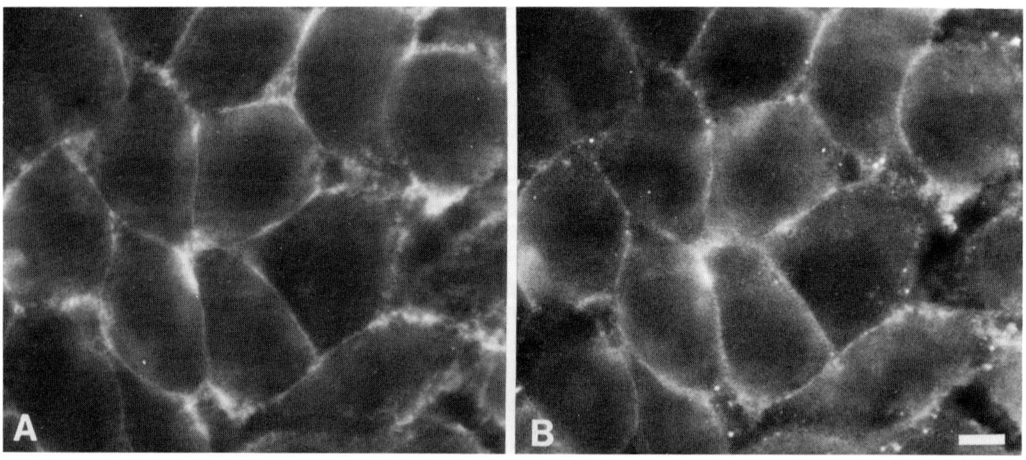

Fig. 3.8. Colocalization of desmoglein 1 and plakoglobin stably expressed in L-cells. Mouse L-cells were cotransfected with an expression construct encoding human plakoglobin cDNA in association with the neomycin resistance marker as well as a construct encoding human desmoglein cDNA. Isolated clones were analyzed by immunofluorescence microscopy for the expression of desmoglein 1 using human autoantibodies from pemphigus foliaceus patients (a generous gift from Dr. J. Stanley) or with a mouse monoclonal against plakoglobin (11E4; a generous gift from Dr. M. Wheelock). Bar = 2 μm.

protein-protein interactions that occurs during desmosome assembly.

FUTURE PROSPECTS

Our understanding of desmosome structure and the interactions among individual desmosome components has been enhanced in recent years by the identification and characterization of cDNAs encoding major desmosomal proteins and their subsequent use in molecular genetic approaches such as those described in this chapter. However, much has yet to be learned about the precise mechanism by which these components associate during desmosome assembly. Furthermore, the precise mechanisms by which desmosome assembly and dissolution are regulated during processes such as wound healing and tumor cell invasion remain to be elucidated. Desmosome assembly is likely to involve a series of sequential interactions, some of which may be transient. Ultimately these interactions must be characterized at the biochemical level. Information gained from molecular genetic experiments such as those described here will provide a valuable guide for future in vitro biochemical studies that will identify direct protein-protein interactions. Furthermore, recent technical advances such as the yeast dihybrid system[71] will provide powerful alternative approaches to confirm suspected protein interactions and uncover interactions that have not been previously predicted. In light of recent advances the next several years promise many new insights into desmosome assembly, regulation and function.

NOTE ADDED IN PROOF

Since submission of this chapter, Band 6, like plakoglobin, has been demonstrated to belong to the *armadillo* multigene family.[74] Additional characterization of the connexin 32-desmoglein chimeras and mapping of the plakoglobin binding site on desmoglein have also been reported.[75]

ACKNOWLEDGMENTS

The authors would like to thank all those who have contributed antibodies, cell lines and other reagents that have been used for experiments from our laboratory including: Orest Blaschuk, Pamela Cowin, Joe Gall, John Stanley, Peter Steinert, Margaret Wheelock and John Stanley. Thanks also go to Laura Nilles for expert assistance with figures and J. Lamb, C. Corcoran, S. Norvell, H. Luu and D.A.D. Parry for their contributions to various aspects of this chapter. Work from the authors' laboratory is supported by grants from the NIH (AR41836), The American Cancer Society (CB-110A and a Faculty Research Award), The March of Dimes Birth Defects Foundation, The Council for Tobacco Research USA, Inc. (#2432A) and a Focused Giving Award from Johnson and Johnson Co.

REFERENCES

1. Luna EJ, Hitt AL. Cytoskeleton-plasma membrane interactions. Science 1992; 258:955-964.
2. Schwarz MA, Owaribe K, Kartenbeck J et al. Desmosomes and hemidesmosomes: constitutive molecular components. In: Palade GE, Alberts BM, Spudich JA, ed. Ann Rev Cell Biol. Palo Alto: Annual Reviews Inc, 1990:461-491.
3. Epstein EH. Molecular genetics of epidermolysis bullosa. Science 1992; 256:799-804.
4. Steinert PM. Structure, function and dynamics of keratin intermediate filaments. J Invest Derm 1993; 100:729-734.
5. Fuchs E. Intermediate filaments and disease: mutations that cripple cell strength. J Cell Biol 1994; 125:511-516.
6. Martinez IR. Fine structural studies of migrating epidermal cells following incision wound. In: Maibach HI, Rovee DT, ed. Epidermal Wound Healing. Chicago: Year Book, 1972: 323-342.
7. Krawczyk WS, Wilgram GF. Hemidesmosome and desmosome morphogenesis during epidermal wound healing. J Ultrastr Res 1973; 45:93-101.
8. Stenn KS, Depalma L. Re-epithelialization. In: Clark RAF, Henson RM, ed. The Molecular and Cellular Biology of Wound Repair. New York: Plenum Press, 1988:321-335.

9. Alroy J, Pauli BU, Weinstein RS. Correlation between numbers of desmosomes and the aggressiveness of transitional cell carcinoma in human urinary bladder. Cancer 1981; 47:104-112.
10. Collins JE, Taylor I, Garrod DR. A study of desmosomes in colorectal carcinoma. Br J Cancer 1990; 62:796-805.
11. Green KJ, Stappenbeck TS, Noguchi S et al. Desmoplakin expression and distribution in cultured rat bladder epithelial cells of varying tumorigenic potential. Exp Cell Res 1991; 193:134-143.
12. Boyer B, Thiery JP. Cyclic AMP distinguishes between two functions of acidic FGF in a rat bladder carcinoma cell line. J Cell Biol 1993; 120:767-776.
13. Farquhar MG, Palade GE. Junctional complexes in various epithelia. J Cell Biol 1963; 17:375-412.
14. Staehelin LA. Structure and function of intercellular junctions. Int Rev Cytol 1974; 39:191-283.
15. Hull BE, Staehelin LA. The terminal web: a reevaluation of its structure and function. J Cell Biol 1979; 81:67-82.
16. Arnn J, Staehelin LA. The structure and function of spot desmosomes. Dermatology 1981; 20:330-339.
17. Cowin P, Kapprell H, Franke WW. The complement of desmosomal plaque proteins in different cell types. J Cell Biol 1985; 101:1442-1454.
18. Steinberg MS, Shida H, Giudice GJ et al. On the molecular organization, diversity and functions of desmosomal proteins. In: Bock G, Clark S, ed. Junctional complexes of epithelial cells. Chichester: John Wiley & Sons, 1987:3-19.
19. Garrod DR. Desmosomes and hemidesmosomes. Curr Opin Cell Biol 1993; 5:30-40.
20. Buxton RS, Cowin P, Franke WW et al. Nomenclature of the desmosomal cadherins. J Cell Biol 1993; 121:481-483.
21. Buxton RS, Magee AI. Structure and interactions of desmosomal and other cadherins. Sem Cell Biol 1992; 3:157-167.
22. Skerrow CJ, Matoltsy AG. Chemical characterization of isolated epidermal desmosomes. J Cell Biol 1974; 63:524-530.
23. Skerrow CJ, Matoltsy AG. Isolation of epidermal desmosomes. J Cell Biol 1974; 63:515-523.
24. Fairley JA, Scott GA, Jensen KD et al. Characterization of keratocalmin, a calmodulin-binding protein from human epidermis. J Clin Invest 1991; 88:315-322.
25. Tsukita S, Tsukita S. Desmocalmin: a calmodulin-binding high molecular weight protein isolated from desmosomes. J Cell Biol 1985; 101:2070-2080.
26. Kapprell H, Owaribe K, Franke WW. Identification of a basic protein of Mr 75,000 as an accessory desmosomal plaque protein in stratified and complex epithelia. J Cell Biol 1988; 106:1679-1691.
27. Cartaud A, Ludosky MA, Courvalin JC et al. A protein antigenically related to nuclear lamin B mediates the association of intermediate filaments with desmosomes. J Cell Biol 1990; 111:581-588.
28. Ouyang P, Sugrue SP. Identification of an epithelial protein related to the desmosome and intermediate filament network. J Cell Biol 1992; 118:1477-1488.
29. Foisner R, Feldman B, Sander L et al. Monoclonal antibody mapping of structural and functional plectin epitopes. J Cell Biol 1991; 112:397-405.
30. Skalli O, Jones JCR, Gagescu R et al. IFAP 300 is common to desmosomes and hemidesmosomes and is a possible linker of intermediate filaments to these junctions. J Cell Biol 1994; 125:159-170.
31. Miller K, Mattey D, Measures H et al. Localisation of the protein and glycoprotein components of bovine nasal epithelial desmosomes by immunoelectron microscopy. EMBO J 1987; 6:885-889.
32. Green KJ, Parry DAD, Steinert PM et al. Structure of the human desmoplakins: implications for function in the desmosomal plaque. J Biol Chem 1990; 265:2603-2612.
33. Green KJ, Stappenbeck TS, Parry DAD et al. Structure of desmoplakin and its association with intermediate filaments. J Dermatol 1992; 19:765-769.
34. Virata MLA, Wagner RM, Parry DAD et al. Molecular structure of the human desmoplakin I and II amino terminus. Proc Natl Acad Sci 1992; 89:544-548.

35. Stappenbeck TS, Bornslaeger EA, Corcoran CM et al. Functional analysis of desmoplakin domains: specification of the interaction with keratin versus vimentin intermediate filament networks. J Cell Biol 1993; 123:691-705.
36. Virata MLA. Identification and characterization of the cDNAs encoding the human desmoplakins I and II, major plaque components of the desmosome involved in intermediate filament organization. Northwestern University, Ph D Thesis 1993; 1-194.
37. Angst BD, Nilles LA, Green KJ. Desmoplakin II expression is not restricted to stratified epithelia. J Cell Sci 1990; 97:247-257.
38. Green KJ, Virata MLA, Elgart GW et al. Comparative structural analysis of desmoplakin, bullous pemphigoid antigen and plectin: members of a new gene family involved in organization of intermediate filaments. Int J Biol Macromol 1992; 14:145-153.
39. O'Keefe EJ, Erickson HP, Bennett V. Desmoplakin I and desmoplakin II: purification and characterization. J Biol Chem 1989; 264:8310-8318.
40. Kelly DE, Kuda AM. Traversing filaments in desmosomal and hemidesmosomal attachments: freeze fracture approaches toward their characterization. Anat Rec 1981; 199:1-14.
41. Stanley JR, Tanaka T, Mueller S et al. Isolation of complementary DNA for Bullous pemphigoid antigen by use of patients' autoantibodies. J Clin Invest 1988; 82:1864-1870.
42. Amagai M, Elgart GW, Klaus-Kovtun V et al. Southern analysis of the 230-kD bullous pemphigoid antigen gene in normal humans, animals, and patients with junctional epidermolysis bullosa. J Invest Dermatol 1991; 97:249-253.
43. Sawamura D, Li K, Chu M et al. Human bullous pemphigoid antigen (BPAG1). J Biol Chem 1991; 266:17784-17790.
44. Wiche G, Becker B, Luber K et al. Cloning and sequencing of rat plectin indicates a 466-kD polypeptide chain with a three-domain structure based on a central alpha-helical coiled coil. J Cell Biol 1991; 114:83-99.
45. Schmelz M, Franke WW. Complexus adhaerentes, a new group of desmoplakin-containing junctions in endothelial cells: The syndesmos connecting retothelial cells of lymph nodes. Eur J Cell Biol 1993; 61:274-289.
46. Pasdar M, Krzeminski KA, Nelson WJ. Regulation of desmosome assembly in MDCK epithelial cells: coordination of membrane core and cytoplasmic plaque domain assembly at the plasma membrane. J Cell Biol 1991; 113:645-655.
47. Stappenbeck TS, Green KJ. The desmoplakin carboxyl terminus coaligns with and specifically disrupts intermediate filament networks when expressed in cultured cells. J Cell Biol 1992; 116:1197-1209.
48. Munro S, Pelham HRB. A C-terminal signal prevents secretion of luminal ER proteins. Cell 1987; 48:899-907.
49. Wiche G, Gromov D, Donovan A et al. Expression of plectin mutant cDNA in cultured cells indicates a role of COOH-terminal domain in intermediate filament association. J Cell Biol 1993; 121:607-619.
50. Kartenbeck J, Schwechheimer K, Moll R et al. Attachment of vimentin filaments to desmosomal plaques in human meningiomal cells and arachnoidal tissue. J Cell Biol 1984; 98:1072-1081.
51. Parrish EP, Marston JE, Mattey DL et al. Size heterogeneity, phosphorylation and transmembrane organisation of desmosomal glycoproteins 2 and 3 (desmocollins) in MDCK cells. J Cell Sci 1990; 96:239-248.
52. Piepenhagen PA, Nelson WJ. Defining E-cadherin-associated protein complexes in epithelial cells: plakoglobin, β- and γ-catenin are distinct components. J Cell Sci 1993; 104:751-762.
53. Shibamoto S, Hayakawa M, Takeuchi K et al. Tyrosine phosphorylation of β-catenin and plakoglobin enhanced by hepatocyte growth factor and epidermal growth factor in human carcinoma cells. Cell Adh Commun 1994; 1:295-305.
54. Nagao S, Kitajima Y, Nagat K et al. Correlation between cell-cell contact formation and activation of protein kinase C in a

human squamous cell carcinoma cell line. J Invest Derm 1989; 92:175-178.

55. Sheu H, Kitajima Y, Yaoita H. Involvement of protein kinase C in translocation of desmoplakins from cytosol to plasma membrane during desmosome formation in human squamous cell carcinoma cells grown in low to normal calcium concentration. Exp Cell Res 1989; 185:176-190.
56. Jones JCR, Goldman RD. Intermediate filaments and the initiation of desmosome assembly. J Cell Biol 1985; 101:506-517.
57. Pasdar M, Nelson WJ. Kinetics of desmosome assembly in Madin-Darby canine kidney epithelial cells: temporal and spatial regulation of desmoplakin organization and stabilization upon cell-cell contact I. Biochemical analysis J Cell Biol 1988; 106:677-685.
58. Pasdar M, Nelson WJ. Kinetics of desmosome assembly in Madin-Darby canine kidney cells: temporal and spatial regulation of desmoplakin organization and stabilization upon cell-cell contact. II. Morphological analysis. J Cell Biol 1988; 106:687-695.
59. Trevor KT, Steben LS. Distribution of desmosomal proteins in F9 embryonal carcinoma cells and epithelial cell derivatives. J Cell Sci 1992; 103:69-80.
60. Bologna M, Allen R, Dulbecco R. Organization of cytokeratin bundles by desmosomes in rat mammary cells. J Cell Biol 1986; 102:560-567.
61. Mattey DL, Garrod DR. Calcium-induced desmosome formation in cultured kidney epithelial cells. J Cell Sci 1986; 85:95-111.
62. Duden R, Franke WW. Organization of desmosomal plaque proteins in cells growing at low calcium concentrations. J Cell Biol 1988; 107:1049-1063.
63. Holm PK, Hansen SH, Sandvig K et al. Endocytosis of desmosomal plaques depends on intact actin filaments and leads to a nondegradative compartment. Eur J Cell Biol 1993; 62:362-371.
64. Burdett IDJ. Internalisation of desmosomes and their entry into the endocytic pathway via late endosomes in MDCK cells. J Cell Sci 1993; 106:1115-1130.
65. Vassar R, Coulombe PA, Degenstein L et al. Mutant keratin expression in transgenic mice causes marked abnormalities resembling a human genetic skin disease. Cell 1991; 64:365-380.
66. Baribault H, Oshima RG. Polarized and functional epithelia can form after the targeted inactivation of both mouse keratin 8 alleles. J Cell Biol 1991; 115:1675-1684.
67. Baribault H, Price J, Miyai K et al. Mid-gestational lethality in mice lacking keratin 8. Genes Devel 1993; 7:1191-1201.
68. Troyanovsky SM, Eshkind LG, Troyanovsky RB et al. Contributions of cytoplasmic domains of desmosomal cadherins to desmosome assembly and intermediate filament anchorage. Cell 1993; 72:561-574.
69. Korman NJ, Eyre R, Klaus-Kortun V et al. Demonstration of an adhering junction molecule (plakoglobin) in the autoantigen of pemphigus foliaceus and pemphigus vulgaris. N Eng J Med. 1989; 321:631-635.
70. Mathur M, Goodwin L, Cowin P. Interactions of the cytoplasmic domain of the desmosomal cadherin Dsg1 with plakoglobin. J Biol Chem 1994; 269:14075-14080.
71. Guarente L. Strategies for the identification of interacting proteins. Proc Natl Acad Sci 1993; 90:1639-1641.
72. Stappenbeck, et al. J Biol Chem, 1994;. In press.
73. Kowalczyk, et al. J Biol Chem, 1994. In press.
74. Hatzfield et al. J Cell Sci, 1994; 107:2259-2270.
75. Troyanovsky, et al. J Cell Biol, 1994; 127:151-160.

CHAPTER 4

PLAKOGLOBIN

Pamela Cowin

PLAKOGLOBIN: THE UNIVERSAL COMPONENT OF ADHESIVE CELL-CELL JUNCTIONS

Desmosomes *(maculae adherentes)* and adherens junctions (*zonulae* and *fasciae adherentes*) are the two major forms of membrane specialization used by cells to attach to each other. Superficially, these two types of junctions resemble one another; both have a widened intercellular space and a pair of electron-dense submembranous mats that associate with the filamentous cytoskeleton. They may be distinguished from each other, however, by subtle variations in these features. Desmosomes have a midline bisecting their intercellular space, whereas adherens junctions are relatively amorphous in this region. Moreover, desmosomes have dense, rigid cytoplasmic plaques underlying the plasma membrane into which intermediate filaments insert, whereas adherens junctions have diffuse, flexible submembranous mats that associate with the actin cytoskeleton.[1]

Some components of these junctions serve similar roles and consequently share certain sequence features. For example, the adhesive function of both junctions is mediated by members of the cadherin superfamily of calcium-dependent cell-cell adhesion proteins. Adherens junctions contain a single member of the classical cadherins, such as E- or N-cadherin, whereas desmosomes are formed by mixtures of members of the desmocollin and desmoglein subsets of the cadherin superfamily. These different types of cadherins are similar in the sequence and repetitive organization of their ectodomains but differ considerably in their cytoplasmic region.[2-9] The divergence of these proteins reflects the necessity of desmosomes and adherens junctions to interact with different elements of the cytoskeleton. This is ultimately achieved through a molecularly distinct set of intermediary proteins that reside in the plaque structure.

A large number of proteins have been localized to the cytoplasmic mats of adherens junctions, or otherwise found to associate with classical cadherins. These include: α, β and γ-catenin, vinculin, radixin, zyxin,

Molecular Biology of Desmosomes and Hemidesmosomes, edited by Jane E. Collins and David R. Garrod.

p120 and several members of the src family of tyrosine kinases.[10-15] Similarly, a plethora of proteins have been localized to the desmosome, such as: desmoplakins I and II, desmocalmin, desmoyokin, band 6, IFAP-300, plectin, 170 kD and a 140 kD lamin-B like protein.[16-22] Of these numerous proteins, only one, called plakoglobin, is shared by both desmosomes and adherens junctions.[23] Thus, plakoglobin is uniquely positioned to play a pivotal role in coordinating the adhesive interactions of cells with its neighbors.

HISTORICAL OVERVIEW

Plakoglobin (Mr~83,000) first appeared in the literature in 1974 when it was referred to as band 5 protein. This name derived from its relative position, following SDS-PAGE, among the eight major polypeptides of subcellular fractions enriched in epidermal desmosomes.[24] Two factors pointed to the likelihood that band 5 protein was a component of the rigid cytoplasmic plaques of desmosomes. First, it was reduced in biochemical preparations that enriched for the desmosomal adhesive core and correspondingly depleted the cytoplasmic plaque structure.[25] Second, band 5 protein was non-glycosylated and therefore thought unlikely to span the plasma membrane.[26]

Immunoelectron-microscopy studies localized band 5 protein to the plaques of desmosomes from tissues rich in these structures, such as tongue and epidermis.[26] Initial surveys of the tissue distribution by immunofluorescence microscopy also described a pattern that was compatible with the presence of band 5 protein in a variety of desmosomes.[27,28] This component was found in abundance in the plasma membrane of stratified epithelial cells; was distributed in a polarized fashion demarcating the terminal bar complex at the border of the apical and basolateral surfaces of simple epithelia cells, and was also present within the junctional complexes in heart known as intercalated discs. Immunoblotting revealed a polypeptide of identical size and isoelectric point to band 5 in each of these tissues.[16] This conservation of biochemical and immunochemical properties was similar to findings for other major plaque proteins, such as desmoplakin I, but contrasted with data on desmosomal and classical cadherins which showed considerable variation in different tissues.[16,29-31]

Subsequent immunochemical studies, however, detected band 5 protein in a number of extra-desmosomal locations. For example, a significant amount of immunofluorescence was observed in the cytoplasm of both cultured cells and tissues resulting from the presence of a globular, dimeric form of band 5 protein.[23,32] Additional membrane-bound forms of band 5 protein were also detected. For example, an antiserum recognizing a mixture of bands 5 and 6 stained adherens junctions in retinal pigmented epithelial cells, suggesting the occurrence of one or both of these antigens at these sites.[33] Monoclonal antibodies detected the presence of band 5 protein in the adherens junctions of polarized epithelial cells and also in a number of specialized cell contacts that are found in lens cells, sertoli cells and endothelia.[23] In contrast, no immunofluorescence was found with monoclonal antibodies at the basal surface of the cells or epithelial tissues indicating that band 5 protein was not present in focal contacts or hemidesmosomes.[23] On the basis of the predicted globular structure of the soluble form of band 5 protein and its localization to the cytoplasmic plaques of all intercellular adhesive junctions, the name plakoglobin was adopted.[23]

THE PLAKOGLOBIN FAMILY

Plakoglobin was cloned and sequenced in 1989 and at the time revealed a proline-rich protein with no significant homology to other proteins in the sequence data bases.[34] Around the same time, the sequence of a segment polarity gene called armadillo was entered into the data base although the close homology between these proteins was not discovered for over two years.[35] These two proteins have 63% identity in their protein sequence and share a similar organization of their central region comprising twelve and a half repeats.[35,36] Recently, β-catenin was also found to share significant sequence similarity to plakoglobin (~63%) and to Armadillo (~70%) and preliminary data suggests that band 6 protein, another desmosomal component, is a member of this family.[37,38]

Each of the four proteins described above associates with adhesive cell-cell junctions. Recently, a growing number of proteins of diverse and intriguing functions, have been described which have low levels of overall sequence identity, but nevertheless comply with a consensus sequence derived from comparisons of the repeat units within the sequences of plakoglobin, β-catenin and armadillo.[39] These include a putative substrate of src called p120, a nucleotide exchange factor GDS, a tumor suppressor gene APC and a suppressor of RNA polymerase I mutations in yeast, SRP1.

PLAKOGLOBIN ASSOCIATION WITH CADHERINS

The pattern of membrane localization of plakoglobin correlates absolutely with regions of cell-cell contact and the presence at these sites of different types of cadherin. Cadherins mediate a number of roles in the cell, in addition to calcium-dependent cell-cell adhesion. For example, they are responsible for cell recognition events during development and their expression produces dramatic changes in cell behavior, morphology and growth.[40] A growing body of evidence indicates that the adhesive properties of cadherins and the cellular responses elicited by cadherin engagement require the cadherin cytoplasmic domain and several associated proteins.[10,41-44]

Association between plakoglobin and desmogleins (DSG), as well as between plakoglobin and classical cadherin-catenin complexes, has been demonstrated by co-precipitation experiments.[45-48] In the case of desmosomes, the association between plakoglobin and several members of the desmoglein family appears to be direct. Patients with the blistering disorders known as pemphigus foliaceus and pemphigus vulgaris contain antibodies within their sera which recognize the desmogleins DSG1 and DSG3 respectively. These sera co-precipitate plakoglobin from Triton X-100 soluble fractions of keratinocytes by virtue of the complexes it forms with both DSG1 and DSG3.[45] It has been postulated, based on the observation that reducing agents disrupt these complexes, that disulfide bonds serve to attach plakoglobin to a cysteine rich sequence found close to the cytoplasmic surface of the plasma membrane of desmogleins.[45,49] This mechanism, however, is unlikely to be used in the intact cell because the cytosol is a reducing environment unfavorable to the formation of such bonds and disulfide isomerase, which catalyses these reactions, is sequestered within the lumen of the endoplasmic reticulum.[50]

Recently, blot-overlay techniques, which have proven successful in dissecting the molecular architecture of the focal contact, have been employed as an alternative method to investigate the interaction between desmogleins and plakoglobin. Desmogleins have a long and complex cytoplasmic region encoded by four exons (see Fig. 4.1).[51] The first exon of this domain encodes a cysteine rich sequence. The sequences encoded by the second and third exons display no conspicuous features. The fourth and last exon encodes a sequence similar to the carboxyl-terminus of classical cadherins followed by a proline rich region, four 30 amino acid repeats and a glycine rich carboxyl-terminus. To define regions important for plakoglobin association, deletion mutants of the DSG1 cytoplasmic domain were created, expressed as fusion proteins and tested for their ability to bind radioiodinated plakoglobin. In this assay, plakoglobin bound to the full length cytoplasmic domain of DSG1 and to several mutants that were progressively deleted from the carboxyl terminus. Plakoglobin binding was lost when sequences near the beginning of exon 4 were removed.[52] The 19 amino acid sequence of the DSG1 cytoplasmic domain, defined by these techniques

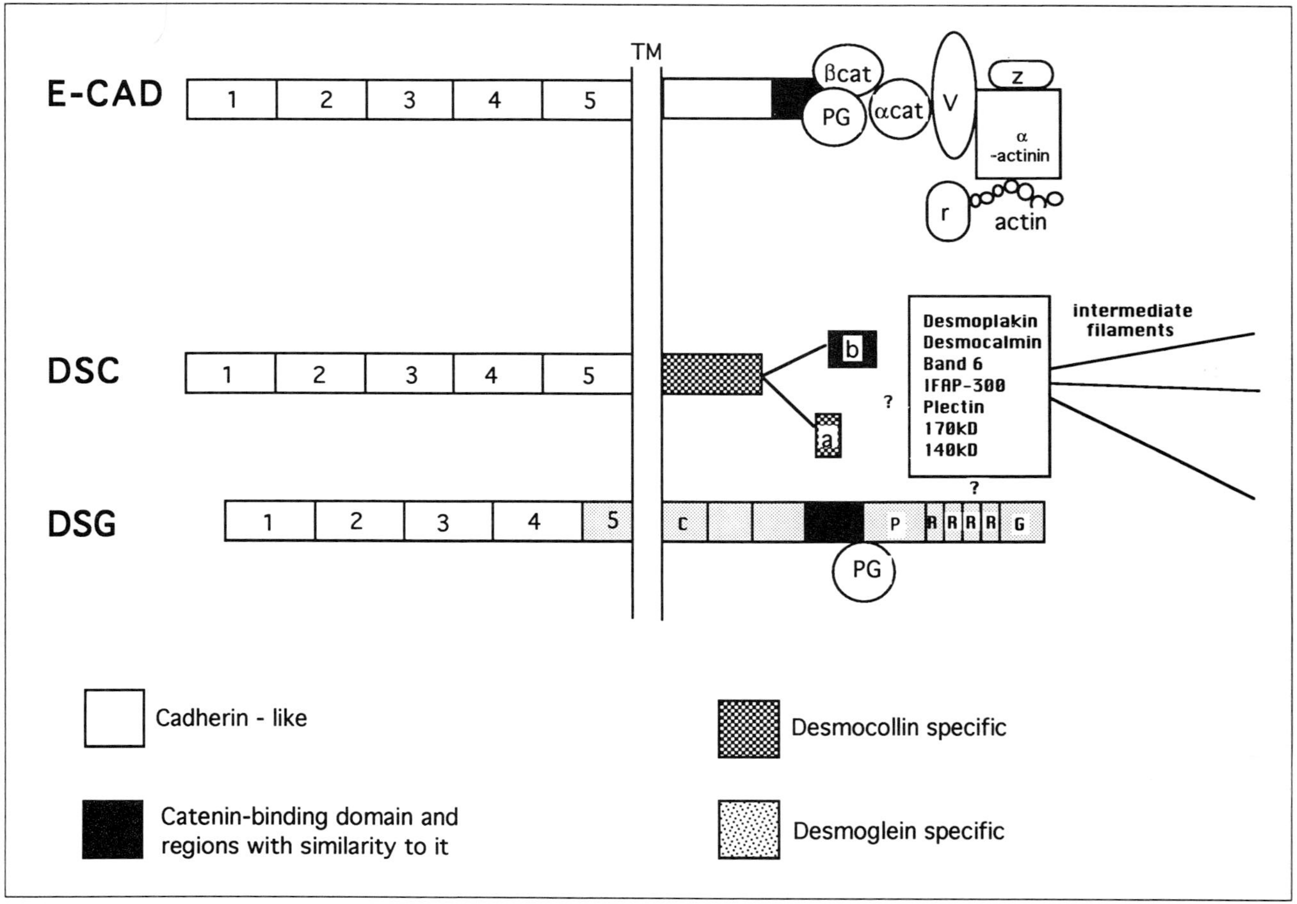

Fig. 4.1. Diagram showing the cytoplasmic interactions of E-cadherin, desmocollins and desmogleins respectively. The members of the protein complex that associates with the cytoplasmic domain of E-cadherin include plakoglobin (PG), b-catenin (bcat), a-catenin (acat), vinculin (v), zyxin (z), a-actinin, actin and radixin (r). Plakoglobin is also known to associate with the cytoplasmic domain of desmoglein. The proteins that are proposed to mediate attachment of intermediate filaments to the desmosomal glycoproteins include desmoplakin, desmocalmin, band 6, 170 kD protein, IFAP 300 and plectin. Numericals 1 to 5 depict the extracellular repeats of the respective proteins. TM = transmembrane domain. P = proline rich region. G = glycine rich region. C = cysteine rich region. a and b = alternatively spliced forms of desmocollins.

as important for binding plakoglobin, reveals strong similarity to corresponding sequences within DSG3, another desmoglein that interacts with plakoglobin. In contrast this sequence reveals little identity to the corresponding sequence of DSC1a, which so far has not been reported to interact with plakoglobin.

Significantly, this 19 amino acid sequence within DSG1 shows considerable similarity to the last ten amino acids of the highly conserved cytoplasmic domain of classical cadherins, suggesting that these cadherins may also bind directly to plakoglobin.[52] Plakoglobin has been co-precipitated from Triton X-100 soluble subcellular fractions of cultured cells by antibodies directed against several different classical cadherins, including those recognizing E-cadherin and N-cadherin.[37,46,48] These antibodies, however, precipitate a large complex including several proteins with which plakoglobin was initially confused, termed catenins. Recently, it has been shown that plakoglobin is distinct from the catenins and furthermore, that classical cadherins form mutually exclusive associations with plakoglobin and β-catenin. This observation eliminates the possibility that plakoglobin binds to cadherins indirectly through association with β-catenin.[53] Steric hindrance of plakoglobin binding by β-catenin and vice versa is consistent with, and would be predicted from, the close proximity of their putative binding sites on E-cadherin.[52] Therefore, at the present time the available data favors a model in which plakoglobin binds directly to both classical cadherins and desmogleins through association with a conserved sequence within their cytoplasmic domains.

Conservation of this binding mechanism by the different types of cadherins would account for the cellular localization of plakoglobin at both adherens junctions and desmosomes. It may be argued, however, that plakoglobin binds to desmosomes and adherens junctions with different biochemical characteristics; plakoglobin binds weakly to the triton X-100 fraction of E-cadherin, yet tenaciously to desmosomes, as evidenced by its association throughout the harsh biochemical extraction conditions used to isolate this type of junction. In this comparison, however, complexes between triton X-100 insoluble E-cadherin and plakoglobin are not considered and similarly the stability of plakoglobin in the fully formed desmosome may depend upon its association with other components in addition to desmoglein. Nevertheless, preliminary experiments in which blot overlay assays of plakoglobin binding have been performed on DSG1 protein deleted from the aminoterminus suggest that additional sequences in the carboxyl terminal region of desmogleins that are not present in classical cadherins may contribute to increased stability of association with this type of cadherin (Goodwin and Cowin, unpublished observations).

RELATIONSHIP OF PLAKOGLOBIN TO THE CATENINS

Several observations indicate that the cytoplasmic sequences of cadherins and the proteins with which they interact regulate the adhesive function of the ectodomain. Firstly, deletion of the last exon leads to a loss in classical cadherins' ability to associate with the actin cytoskeleton, to cluster into junctional sites and to bring about cell-cell adhesion.[10,41-43] Secondly, when the same cytoplasmic domain was overexpressed in *Xenopus* embryos, it produced a dominant negative effect on cell-cell adhesion.[54] At the time these experiments were performed, the protein encoded by the last exon of the cytoplasmic domain of classical cadherins was known to interact with catenins and, therefore, the effects on cell adhesion were attributed to interference with the association of these cytoplasmic regulatory proteins. However, since it is now known that the putative plakoglobin binding site also lies within this same domain of classical cadherins, it is equally plausible that the effects on cadherin function produced by these manipulations may have resulted from disrupting plakoglobin association. The results of recent studies on the expression of the desmoglein cytoplasmic region support this possibility. Overexpression of the DSG1 cytoplasmic domain, which contains a binding site for plakoglobin but does not associate with catenins, produces a dominant negative effect on both desmosome and adherens junction formation in cultured cells.[55] Significantly, expression of deleted forms of this construct in which the region containing the plakoglobin binding site is removed leads to loss of this dominant negative effect on cell junction formation.[38] These experiments clearly indicate the essential role played by plakoglobin in junction assembly.

At the present time the precise mechanism by which cadherin-associated cytoplasmic proteins, including plakoglobin, act is not understood. One interpretation of the deletion and overexpression of cadherin cytoplasmic domains is that catenins/plakoglobin take part in a chain of proteins that anchor cadherins to the cytoskeleton. In this model, the catenin/plakoglobin function is seen as facilitating the adhesive capabilities of cadherins by clustering them into junctions, in a manner analogous to actin-mediated capping of cell surface receptors, thereby creating a site of high adhesive affinity. Several observations make it likely that α-catenin plays this role. α-catenin associates with cadherins late during the biosynthetic process, around the time they reach the cell surface, and becomes post-transcriptionally stabilized in the process.[56] The importance of α-catenin in potentiating the adhesive properties of cadherins has been shown in a lung carcinoma cell line, PC9, which expresses E-cadherin but has no detectable α-catenin.[57] These cells grow as loose aggregates but can be induced to form strongly adherent polarized cysts by transfection and expression of α-catenin. While this experiment does not address the precise role played by α-catenin in the process, the sequence of this protein gives clues as to its likely function.[56,58] α-catenin contains sequence similarity to the self-association domain of vinculin, a plaque protein that is tethered to the cytoskeleton via its interaction with the actin binding protein α actinin. The presence of such a domain within α-catenin suggests that it may cause cadherins to cluster laterally by its self-interaction, or connect them vertically by association with vinculin and hence through a chain of proteins attach them to filamentous actin.[56,58]

The localization of plakoglobin to both desmosomes and adherens junctions, which bind intermediate filaments and filamentous actin respectively, suggests that it is not involved in specifying the type of filament associated with each junction. Furthermore, its absence from cell-substratum junctions such as focal contacts and hemidesmosomes, which share some plaque proteins and engage the same filaments as their cell-cell junction counterparts, further decreases the likelihood that it plays any direct physical role in attaching filaments to the plasma membrane.[23] Indeed, compelling evidence suggests that the well defined actin-α-actinin-vinculin complex binds to focal contacts via talin and, as described above, to adherens junctions via α-catenin.[59] Numerous proteins have been postulated to bind intermediate filaments to desmosomes and hemidesmosomes. Presently, the most likely contenders are desmoplakins I and II, plectin and bullous pemphigoid antigen (BPAG1 or BP230, see chapters 5 and 7), which share a region of sequence identity that is present in truncated forms of desmoplakins that have been shown to bind to and disrupt intermediate filaments[60] (see chapter 3).

An alternative view of the function of cadherin-associated proteins is that they may take part in signaling cascades that promote junction assembly and/or are involved in the multiple cellular responses to cadherin engagement.[10,41,42] Several studies to be discussed later, suggest that members of the plakoglobin family take part in transducing morphoregulatory signals.[61] One intriguing observation on the sequence similarity among plakoglobin family members hints that they may play a role in conventional signal transduction cascades that could serve to modulate junction assembly and/or transmit further information. β-catenin, plakoglobin and other family members bear a structural resemblance to a protein called GDS.[62,63] GDS activates a number of G-proteins including, Rac, Rho, Rap-1 and Ki-ras by promoting the exchange of GDP for GTP.[64] Several of these G-proteins are intimately involved in the interactions of actin with membranes and have profound effects on cell shape.[65] Since GDS is composed almost entirely of the 42 amino acid repeat units that were first described in armadillo, it is hypothesized that the repeat unit must mediate the nucleotide exchange activity.[39] If so, then it possible that the repeats found in other members of the family may operate in a similar fashion. G-proteins have been shown to play a role in tight junction assembly and have also been implicated as downstream effectors of N-cadherin induced neurite extension.[66]

MODULATION OF CELL ADHESION BY PHOSPHORYLATION OF PLAKOGLOBIN FAMILY MEMBERS

Both plakoglobin and β-catenin associate with cadherins prior to cell adhesion. For example, newly synthesized plakoglobin and β-catenin rapidly associate with the newly synthesized precursor E-cadherin (Witcher and Cowin unpublished observations).[67] Furthermore, all three catenins will associate with forms of cadherin that are adhesion defective due to deletion of their ectodomains.[41] Therefore while previously described experiments have demonstrated that association of these cytoplasmic proteins with cadherins is a prerequisite for junction formation, the formation of this complex is not the factor which triggers this event. Clearly, other regulatory steps, such as phosphorylation, must intervene to initiate cell adhesion and to produce a response to this process.

Several members of the plakoglobin family are reported to be tyrosine phosphorylated, and hence are substrates of conventional signaling molecules. Two studies show a correlation between levels of tyrosine phosphorylation of plakoglobin and β-catenin and poor adhesive behavior in response to growth factor stimulation and transformation. In one study, EGF and HGF were found to induce cell scattering to varying degrees in three different cell lines.[68] These agents, while loosening cell contacts had no effect on the overall levels of cadherins or catenins. However, the degree of cell dispersal correlated with an increase in tyrosine phosphorylation of both β-catenin and plakoglobin. In these and other studies, phosphorylation does not appear to affect the ability of these components to form complexes with cadherins, and it appears that the soluble forms of these proteins are also tyrosine phosphorylated to an equal degree.

In a second study, src and fos transformed fibroblasts were found to aggregate poorly by comparison to the non-transformed parental cell line and displayed a relatively poor ability to sequester their cadherins into cell-cell contact regions.[69] These effects of transformation could be alleviated to some extent by treating the cells with a tyrosine kinase inhibitor, but were aggravated by inhibitors of phosphatases. In each case, the effects on cell aggregation correlated with the levels of tyrosine phosphorylation of β-catenin in the cadherin-catenin complex.[69]

Phosphorylation of other plakoglobin family members has also been correlated with abrogation of cell adhesion in response to src induced transformation. For example, p120, which we have recently localized to a distinct subset of adherens junctions (Murrell and Cowin, unpublished observations), is tyrosine phosphorylated in response to several growth factors and in cells transformed with src.[70] It is not, however, phosphorylated in cells in which src mutants have been introduced that retain kinase ability but are transformation defective. This suggests that the tyrosine phosphorylation of p120 may be a critical event in producing the transformed phenotype associated with RSV infection.[70]

Members of the src family of kinases are enriched at both focal contacts and adherens junctions suggesting that phosphorylation of constituents within these sites occurs in normal cells.[14] pp60src is known to disrupt both of these junctions junctions.[14,71,72] The interactions of pp60src with p125FAK, a constituent of focal contacts are well understood. pp60src constitutively activates p125FAK a kinase that is normally phosphorylated in response to integrin-mediated cell-substratum adhesion. In this manner pp60src mimics an adhesion-induced signaling pathway and thereby facilitates anchorage independent growth of cell.[73] In contrast, the substrates of pp60src in the adherens junction are not known. However, the observation that three members of the plakoglobin family located at these sites become tyrosine phosphorylated in response to both src and growth factors, and that this degree of phosphorylation correlates with impaired adhesion, suggests that phosphorylation and dephosphorylation of such junctional components could play a central role in regulating junctional stability. This is in agreement with the observation that inhibitors of tyrosine kinases prevent the disassembly of cell junctions that normally takes place in response to lowering of calcium levels in the cell culture medium.[74] In addition to tyrosine kinases of the src family that are enriched at cell junctions, a growing number of kinases (cRET) and phosphatases (PTPu) have been reported to contain cadherin-like sequences suggesting that they may localize to sites of cell adhesion and regulate their function.[75,76]

ROLE OF PLAKOGLOBIN-FAMILY MEMBERS IN PATTERN FORMATION

A role for plakoglobin family members in the transduction and/or implementation of patterning cues during development was first suggested by studies on *Armadillo,* the product of a *Drosophila* segment polarity gene.[35,36,77] Segment polarity genes are thought to act in concert to specify patterning within each body segment of the embryo.[78,79] This pattern first becomes evident in the cuticle overlying the epidermis which contains denticles in the anterior half of the segment but is smooth in the posterior half. In current models of this genetic pathway (see Fig. 4.2), activity of a transcription factor, Engrailed (en) is re-

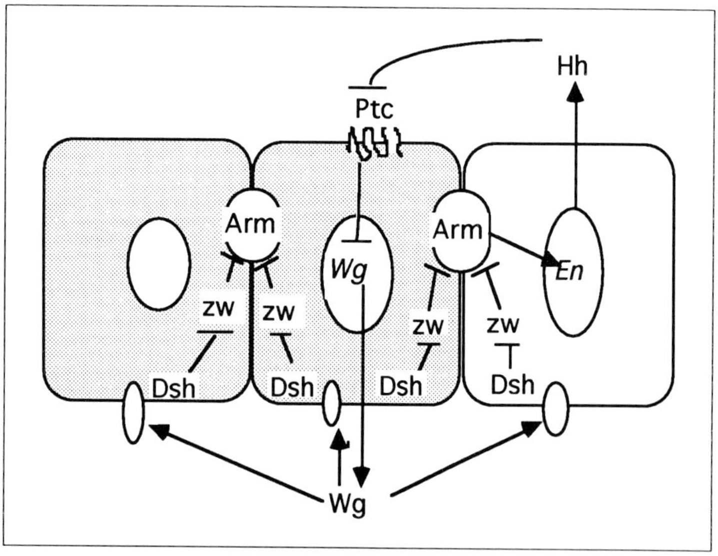

Fig. 4.2. Schematic of the wingless signaling pathway. Engrailed (en) is a transcription factor that is required for expression of a secreted protein Hedgehog (hh) which binds to Patched (ptc), a transmembrane receptor on the surface of neighboring cells and to negate its inhibitory effect on the expression of Wingless (wg) protein. Once secreted Wingless binds to an unknown receptor on the surface of neighboring cells and acts via Di-sheveled (Dsh), a cytoplasmic protein of unknown function, to negatively regulate Zeste White (zw), a serine threonine kinase. In the absence of Zeste White activity, the cytoplasmic pool of Armadillo (arm) is postranscriptionally upregulated (shaded cells) and a critical portion of this protein associates with the membrane. Armadillo *acts upstream of* engrailed, *causing the continued expression of this transcription factor.*

quired for expression of Hedgehog (hh), a protein which is secreted and diffuses across the segment, establishing a morphogenetic gradient. Hedgehog is thought to bind to Patched (ptc), a transmembrane receptor on the surface of neighboring cells, and to negate its inhibitory effect on the expression of a protein called Wingless (wg). Wingless is a secreted protein that acts locally by binding to an unknown receptor on the surface of neighboring cells. It acts via Disheveled (Dsh), a cytoplasmic protein of unknown function, to negatively regulate Zeste White (zw), a serine threonine kinase. In the absence of Zeste White activity, the cytoplasmic pool of Armadillo (arm) is postranscriptionally upregulated and a critical portion of this protein associates with the membrane.[77,80,81] Epistasis experiments show that *armadillo* acts upstream of *engrailed*, causing the continued expression of this transcription factor and hence establishing feed-back on the patterning process.[82]

Several models can be imagined in which the pattern information is implemented by Armadillo. The strong homology among Armadillo, plakoglobin and β-catenin suggests that further transmission involves engagement of the cadherin based cell adhesion system. This is further supported by the finding of Armadillo in a complex with α-catenin and the immunolocalization of Armadillo to structures that resemble adherens junctions.[83,84] In one model, cadherin-mediated adhesion is seen as playing a passive facilitative role in the process of pattern formation. For example, it may simply be required to bring cells close enough for other receptors to engage and transmit morphoregulatory signals. It is well known that cells require strong adhesion to be able to assemble gap junctions which have been shown in many developmental systems to be essential ports through which second messengers bearing positional information pass among a community of adherent cells to coordinate their differentiation.[85,86] Interestingly, injection of *wnt* mRNA has been reported to enhance the formation of gap junctions in *Xenopus* embryos.[86]

In an alternative model, engagement of the cadherin adhesion system could lead directly to phenotypic changes in cell shape and tissue morphology, not only by causing cell to adhere, but also by leading to localized reorganization of both the actin and intermediate filament cytoskeleton. Dramatic changes in cell morphology, following transfection of E-cadherin, are documented in a number of cultured cell lines. For example, fibroblasts can be induced to adopt an epithelial morphology in response to E-cadherin expression which is accompanied by a reorganization and polarization of the subcortcal cytoskeleton and other cell surface receptors.[10,40,44]

In a third model, the pattern information is further transduced by conventional signal cascades resulting from cadherins engagement. Experiments in PC12 cells support this possibility. PC12 cells can be induced to adopt a neuronal phenotype by plating them upon fibroblasts expressing N-cadherin.[66] This striking change in morphology can be inhibited by cholera toxin, calcium channel antagonists, and specific tyrosine kinase inhibitors. Furthermore, PC12 cells could be induced to extend neurites in the absence of cadherin or other CAM activity by directly stimulating this signaling cascade.[87,88]

Finally, it can be argued that the significant step in the *wingless* pathway is the upregulation of the cytosolic pool rather than the membrane-bound form of armadillo.[81] In this model, cytoplasmic armadillo is thought to be a direct messenger of the pattern information; perhaps acting in the transduction machinery in a manner analogous to that described for GDS, or by associating with and potentiating the action of a further component in an analogous fashion to the interaction of β-catenin and APC, which appears to regulate cell growth.

Several studies in vertebrates and in cultured mammalian cells demonstrate that β-catenin and plakoglobin are involved in transmitting developmental cues. For example, injection of monovalent fragments of antibodies directed against β-catenin or mRNA encoding this protein into *Xenopus* embryos produces duplication of the dorsoanterior axis.[89] Significantly, this phenotype is also produced by introduction of mRNAs encoding *Wnt,* a vertebrate homologue of the *Drosophila* gene Wingless.[90] These results further reinforce the possibility that *Wnt* and β-catenin operate in concert to control cell fate.

Two studies in mammalian cultured cell lines further demonstrate that the mechanisms of patterning used in the Wingless pathway in *Drosophila* are highly conserved. The neuropheochromocytoma cell line PC12 cells, which are normally rounded and poorly adhesive, adopt an epithelial appearance, forming tightly adherent flattened colonies of cells when transfected with *Wnt-1*.[91] These changes are accompanied by a post-transcriptional elevation of plakoglobin. A shift is seen in the relative amounts of plakoglobin from a cytosolic fraction to a membrane-bound form and this protein localizes to points of cell-cell contact. This is accompanied by a seven fold increase in E-cadherin expression and a marked increase in the ability of these cells to adhere to one another in aggregation assays.[91] A subsequent study reported similar effects on the levels of plakoglobin in AtT20 cells and C57MG cells in response to Wnt-1 expression.[53] The latter study also noted that these particular *Wnt-1* expressing cells showed an increase in the stability of the β-catenin bound to N-cadherin and hence an increase in the relative steady-state amount of cadherin associated with β-catenin as opposed to that complexed to plakoglobin.

Interestingly, *Wnt-1* produces an epithelial phenotype and contact-inhibited growth in PC12 cells, whereas its effect on C57MG morphology is the converse, causing a change from an epithelial to fibroblastic phenotype and leading to a loss of contact-induced inhibition of cell growth.[91,92] At the present time, it is not clear how *Wnt* induces these opposing morphological changes and responses to cell contact in different cultured cells. One possible mechanism might involve the choice between whether cadherins associate with plakoglobin or β-catenin and the consequences of such a choice for the levels of cytosolic pools of these proteins which appear to play a role in growth regulation.

ROLE OF PLAKOGLOBIN-FAMILY MEMBERS IN TUMOR SUPPRESSION

Several proteins in the plakoglobin family exist as cytosolic dimers in addition to their membrane bound forms. A re-examination of the significance of the cytosolic pools of plakoglobin-like proteins has been provoked by the recent finding of β-catenin in a cytosolic complex with the product of a tumor supressor gene called APC.[93,94] APC is linked to an inherited disease of the colon called familial adenomatous polyposis (FAP) and mutations that result in truncation of its carboxyl-terminal domain have been implicated in a number of sporadic colon carcinomas.[95,96] The product of this gene has low but significant homology to plakoglobin in its amino-terminal third and is known to form cytosolic homodimers through this region. APC binds to β-catenin through a sequence that is repeated three times in its central domain that has no homology to plakoglobin.[93] While most mutations that give rise to colonic polyps delete the carboxyl-terminal domain that does not contain a binding site for β-catenin, it has nevertheless been postulated that these mutated forms of APC have reduced affinity for β-catenin. If this is true, this would place β-catenin in a pathway regulating cell growth and cell adhesion.

Expression studies of deletion mutants in *Drosophila* indicate that the individual repeats in the protein operate autonomously yet in an additive fashion. It is therefore intriguing that β-catenin and APC both contain arm repeats yet they do not use them to interact with each other. This suggests that interaction between these two proteins may in some way serve to amplify or modulate the function of their repeats. This is the first example of two proteins from the plakoglobin family interacting with each other and raises the possibility that the specific functions of plakoglobin-like proteins may be specified by the type of dimer they form in a manner analogous to transcription factors.

CONCLUSION

In summary, plakoglobin and β-catenin interact with the carboxyl termini of cadherins and appear to regulate their adhesive function. Three other related proteins, band 6, Armadillo, and p120 also localize to cell-cell junctions. Plakoglobin, β-catenin, Armadillo and APC are present in a soluble form in the cytoplasm, suggesting that they play additional roles within the cell. Armadillo is required for patterning by *wingless* signals in *Drosophila* and this role has clearly been maintained by plakoglobin and β-catenin in vertebrate cells. The precise mechanism by which these proteins achieve their functions is unclear. Two questions that remain to be solved are: whether the morphological changes, seen in cells and embryos as a result of expressing *Wnt* and *Wingless* respectively, are due to the actions of the cytosolic pools or the membrane bound pools of these proteins? Secondly, whether plakoglobin, β-catenin and Armadillo are upstream facilitators or downstream effectors of cadherin function in the *Wnt/Wingless* signaling pathway? Answers to these questions should be greatly facilitated by the possibility to use *Drosophila* genetics to analyze junction assembly.

The existence of several proteins (p120, APC, GDS, SRP1) with a similar repeat structure to plakoglobin is intriguing. Since p120 localizes to junctional sites and APC forms complexes with β-catenin, the sequence similarity among these proteins, while statistically low, appears to be functionally meaningful. If this is generally true, then the similarity of plakoglobin-like proteins to GDS, a nucleotide exchange factor, suggests they may function in conventional signaling pathways and warrants further investigation of this potential activity and definition of their target proteins.

At the present time, models for the function of plakoglobin-like proteins are purely speculative. If an equilibrium exists between the membrane-bound and cytosolic forms of these proteins and competition arises among them to form complexes with cadherins, then models can be envisaged in which cell adhesion and contact inhibition of growth can be coordinated. For example, a factor such as Wnt which appears to stabilize complexes between β-catenin and cadherin may as a consequence deplete the amount of cytosolic β-catenin available to complex APC. This delicate shift in the equilibrium of the various β-catenin pools might then account for the failure of C57MG cells to inhibit their growth in response to cell contact. In contrast, the upregulation of membrane bound plakoglobin by Wnt in PC12 cells is predicted to displace β-catenin from E-cadherin into the cytosolic fraction, thereby favoring complexes with APC and enhancing contact inhibition of cell growth. While this model is compatible with the existing data, it is but one of many possible scenarios. Clearly many other models can be imagined and testing them promises to be an exciting chapter of future research.

ACKNOWLEDGMENTS

I would like to thank Dr. James Salzer and Gail Skovronsky for helpful comments and the National Institutes of Health (R01-GM47429), The PEW Charitable Trusts, and the Dermatology Foundation, USA for financial support of work contributed to this chapter.

REFERENCES

1. Farquhar MG, Palade GE. Junctional complexes in various epithelia. J. Cell Biol. 1963; 17:375-412.
2. Goodwin L, Hill JE, Raynor K, Raszi L, Manabe M, Cowin P. Desmoglein shows extensive homology to the cadherin family of cell adhesion molecules. Biochem Biophys Res Com 1990; 173:1224-30.
3. Mechanic S, Raynor K, Hill JE, Cowin P. Desmocollins form a novel subset of the

cadherin family of cell adhesion molecules. Proc Natl Acad Sci USA 1991; 88:4476-4480.

4. Collins JE, Legan PK, Kenny TP, MacGarvie J, Holton JL, Garrod DR. Cloning and sequence analysis of desmosomal glycoproteins 2 and 3 (desmocollins): cadherin-like desmosomal adhesion molecules with heterogeneous cytoplasmic domains. J Cell Biol 1991; 113:381-391.
5. Koch PJ, Walsh MJ, Schmelz M, Goldschmidt MD, Zimbelmann R, Franke WW. Identification of desmoglein a constititive desmosomal glycoprotein as a member of the cadherin family of cell adhesion molecules. Eur J Cell Biol 1990; 53:1-12.
6. Koch PJ, Goldschmidt MD, Zimbelmann R, Troyanovsky R, Franke WW. Complexity and expression patterns of the desmosomal cadherins. Proc Natl Acad Sci USA 1992; 89:353-357.
7. Nilles LA, Parry DA, Powers EE, Angst BD, Wagner RM, Green KJ. Structural analysis and expression of human desmoglein: a cadherin-like component of the desmosome. J Cell Science 1991; 99:809-821.
8. Parker AE, Wheeler GN, Arnemann J, Pidsley SC, Ataliotis P, Thomas CL, Rees DA, Magee AI, Buxton RS. Desmosomal glycoproteins II and III. Cadherin-like junctional molecules generated by alternative splicing. J Biol Chem 1991; 266:10438-10445.
9. Wheeler GN, Parker AE, Thomas CL, Ataliotis P, Poynter D, Arnemann J, Rutman AJ, Pidsley SC, Watt FM, Rees DA, al. e. Desmosomal glycoprotein DGI, a component of intercellular desmosome junctions, is related to the cadherin family of cell adhesion molecules. Proc Natl Acad Sci USA 1991; 88:4796-800.
10. Ozawa M, Baribault H, Kemler R. The cytoplasmic domain of the cell adhesion molecule uvomorulin associates with three independent proteins structurally related in different species. EMBO J 1989; 8:1711-1717.
11. Geiger B, Schmid E, Franke WW. Spatial distribution of proteins specific for desmosomes and adhaerens junctions in epithelial cells demonstrated by double immunofluorescence microscopy. Differentiation 1983; 23:189-205.
12. Crawford AW, Michelson JW, Beckerle M. An interaction between zyxin and alpha-actinin. J Cell Biol 1992; 116:1381-1393.
13. Tsukita S, Tsukita S. A new 82kD barbed end capping protein Radixin localized in the cell cell adherens junction-purification and characterization. J Cell Biol 1989; 108:2369-2382.
14. Tsukita S, Oishi K, Akiyama T, Yamanashi Y, Yamamoto T. Specific proto-oncogene tyrosine kinases of a src family are enriched in cell-to-cell adherens junctions where the level of tyrosine phosphorylation is elevated. J Cell Biol 1991; 113:867-879.
15. Tsukita S, Tsukita S. Molecular Linkage between cadherins and actin filaments in cell-cell adherens junctions. Curr Op Cell Biol 1993; 4:834-839.
16. Cowin P, Kapprell HP, Franke WW. The complement of desmosomal plaque proteins in different cell types. J Cell Biol 1985; 101:1442-54.
17. Cautaud A, Ludosky MA, Courvalin JC, Cautaud J. A Protein antigenically related to nuclear Lamin B mediates the association of intermediate filaments with desmosomes. J Cell Biol 1990; 111:581- 588.
18. Kapprell HP, Owaribe K, Franke WW. Identification of a basic protein of Mr 75,000 as an accessory desmosomal plaque protein in stratified and complex epithelia. J Cell Biol 1988; 106:1679-91.
19. Skalli O, Jones JCR, Gagescu R, Goldman RD. IFAP300 is common to desmosomes and hemidesmosomes and is a possible linker of intermediate filaments to these junctions. J Cell Biol 1994; 125:159- 170.
20. Tsukita S, Tsukita S. Desmocalmin: a calmodulin-binding high molecular weight protein isolated from desmosomes. J Cell Biol 1985; 101:2070-80.
21. Wiche G, Krepler R, Artlieb U, Pytela R, Aberer W. Identification of plectin in different human cell types and immunolocalization at epithelial basal cell surface membranes. Exp Cell Res 1984; 155:43-9.
22. Wacker IA, Rickard JE, DeMey JR, Kreis TE. Accumulation of a Microtubule binding Protein, pp170, at Desmosomal Plaques.

J Cell Biol 1992; 117:813-824.
23. Cowin P, Kapprell H-P, Franke WW, Tamkun J, Hynes RO. Plakoglobin: A protein common to different kinds of intercellular adhering junctions. Cell 1986; 46:1063-1073.
24. Skerrow CJaMAG. Chemical characterization of isolated epidermal desmosomes. J Cell Biol 1974; 63:524-531.
25. Gorbsky G, Steinberg MS. Isolation of the intercellular glycoproteins of desmosomes. J Cell Biol 1981; 90:243-248.
26. Gorbsky G, Cohen SM, Shida H, Giudice GJ, Steinberg MS. Isolation of the non-glycosylated proteins of desmosomes and immunolocalization of a third plaque protein: desmoplakin III. Proc Natl Acad Sci USA 1985; 82:810-4.
27. Cowin P, Garrod DR. Antibodies to epithelial desmosomes show wide tissue and species cross-reactivity. Nature 1983; 302:148-150.
28. Cowin P, Mattey D, Garrod DR. Distribution of desmosomal components in tissues of vertebrates, studied by fluorescent antibody staining. J Cell Sci 1984; 66:119-132.
29. Cohen SM, Gorbsky G, Steinberg MS. Immunochemical characterization of related families of glycoproteins in desmosomes. J Biol Chem 1983; 258:2621-2627.
30. Giudice GJ, Cohen SM, Patel NH, Steinberg MS. Immunological comparison of desmosomal components from several bovine tissues. J Cell Biochem 1984; 26:35-45.
31. Suhrbier A, Garrod DR. An investigation of the molecular components of desmosomes in epithelial cells of five vertebrates. J Cell Sci 1986; 81:223-242.
32. Kapprell H-P, Cowin P, Franke WW. Biochemical characterization of the soluble form of the junctional plaque protein, plakoglobin, from different cell types. Eur J Biochem 1987; 166:505-517.
33. Docherty RJ, Edwards JG, Garrod DR, Mattey DL. Chick Embryonic Pigmented Retina is one of the group of epitheliod tissues that lack cytokeratins and desmosomes and have intermediate filaments composed of vimentin. J Cell Sci 1984; 71:61-74.
34. Franke WW, Goldschmidt MD, Zimblemann R, Mueller HM, Schiller DL, Cowin P. Molecular cloning and amino acid sequence of plakoglobin, common junctional plaque protein. Proc Natl Aca Sci USA 1989; 86:4027-4031.
35. Riggleman B, Wieschaus E, Schedl P. Molecular analysis of the armadillo locus: uniformly distributed transcripts and a protein with novel internal repeats are associated with Drosophila segment polarity gene. Genes and Dev 1989; 3:96-113.
36. Peifer M, Weischaus E. The segment polarity gene armadillo encodes a functionally modular protein that is the Drosophila homolog of human plakoglobin. Cell 1990; 63:1167-1178.
37. McCrea PD, Turck CW, Gumbiner B. A homolog of the armadillo protein in Drosophila (plakoglobin) associated with E-cadherin. Science 1991; 254:1359-61.
38. Shafer S, Troyanovsky SM, Heid HW, Eshkind LG, Koch PJ, Franke WW. Cytoskeletal architecture and epithelial differentiation: molecular determinants of cell interaction and cytoskeletal anchorage. CR Acad Sci Paris 1993; 316:1316-1323.
39. Peifer M, Berg S, Reynolds A. A repeating amino acid motif defines a new relationship among proteins with roles in signaling, tumor suppression, cell-cell adhesion and guanine nucleotide exchange. Cell 1994; 76:789-791.
40. Nose A, Nagafuchi A, Takeichi M. Expressed recombinant cadherins mediate cell sorting in model systems. Cell 1988; 54:993-1001.
41. Ozawa M, Ringwald M, Kemler R. Uvomorulin-catenin complex formation is regulated by a specific domain in the cytoplasmic region of the cell adhesion molecule. Proc Natl Acad Sci USA 1990; 87:4246-4250.
42. Nagafuchi A, Takeichi M. Transmembrane control of cadherin-mediated cell adhesion: a 94 kDa protein functionally associated with a specific region of the cytoplasmic domain of E-cadherin. Cell Regulation 1989; 1:37-44.
43. Nagafuchi A, Takeichi M. Cell binding function of E-cadherin is regulated by the cytoplasmic domain. Embo J 1988;

7:3679-3684.
44. McNeil H, Ozawa M, Kemler R, Nelson WJ. Novel function of the cell adhesion molecule uvomorulin as an inducer of cell surface polarity. Cell 1990; 62:309-316.
45. Korman NJ, Eyre RW, Klaus-Kovtun V, Stanley JR. Demonstration of an adhering junction molecule (plakoglobin) in the autoantigens of pemphigus foliaceus and pemphigus vulgaris. N Engl J Med 1989; 321:631-635.
46. Knudsen KA, Wheelock MJ. Plakoglobin, or an 83-kD homologue distinct from beta-catenin, interacts with E-cadherin and N-cadherin. J Cell Biol 1992; 118:671-679.
47. Peifer M, McCrea P, Green KJ, Wieschaus E, Gumbiner BM. The vertebrate adhesive junction proteins beta-catenin and plakoglobin and the Drosophila segment polarity gene armadillo form a multigene family with similar properties. J Cell Biol 1992; 118:681-691.
48. Piepenhagen PA, Nelson WJ. Defining E-cadherin-associated protein complexes in epithelial cells: plakoglobin, beta and gamma catenin are distinct components. J Cell Sci 1993; 104:751-762.
49. Amagai M, Klaus KV, Stanley JR. Autoantibodies against a novel epithelial cadherin in pemphigus vulgaris, a disease of cell adhesion. Cell 1991; 67:869-877.
50. Alberts B, Bray D, Lewis J, Raff M, Roberts K, Watson JD. The Molecular Biology of the Cell. 1989.
51. Puttagunta S, Mathur M, Cowin P. Structure of DSG1, the bovine desmosomal cadherin gene encoding the pemphigus foliaceus antigen. J Biol Chem 1994; 269:1949-1955.
52. Mathur M, Goodwin L, Cowin P. The interactions of a desmosomal cadherin, DSG1, with plakoglobin. J Biol Chem 1994; 269:1-6.
53. Hinck L, Nelson WJ, Papkoff J. Wnt-1 modulates cell-cell adhesion in mammalian cells by stabilizing β-catenin binding to the cell adhesion protein cadherin. J Cell Biol 1994; 124:729-741.
54. Kintner C. Regulation of embryonic cell adhesion by the cadherin cytoplasmic domain. Cell 1992; 69:225-236.
55. Troyanovsky SM, Eshkind LG, Troyanovsky RB, Leube RE, Franke WW. Contributions of cytoplasmic domains of desmosomal cadherins to desmosome assembly and intermadiate filament anchorage. Cell 1993; 72:561-574.
56. Nagafuchi A, Takeichi M, Tsukita S. The 102 kd cadherin-associated protein: similarity to vinculin and posttranscriptional regulation of expression. Cell 1991; 65:849-57.
57. Hirano S, Kimoto N, Shimoyama Y, Hirohashi S, Takeichi M. Identification of a neural alpha-catenin as a key regulator of cadherin function and multicellular organization. Cell 1992; 70:293-301.
58. Herrenknecht K, Ozawa M, Eckerskorn C, Lottspeich F, Lenter M, Kemler R. The uvomorulin-anchorage protein alpha-catenin is a vinculin homologue. Proc Natl Acad Sci 1991; 88:9156-60.
59. Burridge K, Mangeat P. An interaction between Vinculin and Talin. Nature 1984; 308:744-746.
60. Stappenbeck T, Green KJ. The desmoplakin carboxyl terminus coaligns with and specifically disruptsintermediate filament networks when expressed in cultured cells. J Cell Biol 1992; 116:1197-1209.
61. Cowin P, Brown AMC. In: Bernfield M, ed. Molecular Basis of Morphogenesis. New York: Wiley-Liss, Inc, 1993:44-66.
62. Kikuchi A, Kaibuchi K, Hori Y, Nonaka H, Sakoda T, Kawamura M, Mizuno T, Takai Y. Oncogene 1992; 7:289-293.
63. Peifer. Wingless signal and zeste-white 3 kinase trigger opp. Development 1994.
64. Hiraoka K, Kaibuchi K, Ando S, Musha T, Takaishi K, Mizuno A, Menard L, Tomhave E, Didsbury J, Snyderman R, Takai Y. Biochem Biophys Res Commun 1992; 182:921-930.
65. Ridley A, Hall AJ. The small GTP binding protein Rho regulates the assembly of focal adhesions and actin stress fibres in response to growth factors. Cell 1992; 70:389-399.
66. Doherty P, Ashton SV, Moore S, Walsh FS. Morphoregulatory activities of NCAM and N-Cadherin can be accounted for by G protein-dependent activation of L and N type channels. Cell 1991; 67:21-33.

67. Ozawa M, Kemler R. Molecular organization of the uvomorulin- catenin complex. J Cell Biol 1992; 116:989-96.
68. Shibamoto S, Hayakawa M, Takeuchi K, Hori T, Oku N, Miyazawa K, Kitamura M, Takeichi M, Ito F. Tyrosine phosphorylation of beta-catenin and plakoglobin enhanced by hepatocyte growth factor and epidermal growth factor in human carcinoma cells. Cell Adhesion and Communication. 1994; 1:295-305.
69. Matsuyoshi N, Hamaguchi M, Taniguchi S, Nagafuchi A, Tsukita S, Takeichi M. Cadherin-mediated cell-cell adhesion is perturbed by v-src tyrosine phosphorylation in metastatic fibroblasts. J Cell Biol 1992; 118:703-714.
70. Reynolds AB, Rosel DJ, Kanner SB, Parsons JT. Transformation-specific tyrosine phosphorylation of a novel cellular protein in chicken cells expressing oncogenic variants of the avian cellular src gene. Mol Cell Biol 1989; 9:629-638.
71. Volberg T, Geiger B, Dror R, Zick Y. Modulation of intercellular adherens-type junctions and tyrosine phosphorylation of their components in RSV-transformed cultured chick lens cells. Cell Reg 1991; 2:105-120.
72. Warren SL, Nelson WJ. Nonmitogenic morphoregulatory action of pp60v-src on multicellular epithelial structures. Mol Cell Biol 1987; 7:1326-37.
73. Lipfert L, Haimovich B, Schaller MD, Cobb BS, Parsons JT, Brugge JS. Integrin dependent phosphorylation and activation of the protein tyrosine kinase pp125 FAK in platelets. J Cell Biol 1992; 119:905.
74. Citi S. Protein kinase inhibitors prevent junction dissociation induced by low extracellular calcium in MDCK cells. J Cell Biol 1992; 117:169-178.
75. Iwamoto T, Taniguchi M, Asai N, Ohkusu K, Nakashima I. cDNA Cloning of Mouse ret proto-oncogene and its sequence similarity to the cadherin family. Oncogene 1992.
76. Tonks NK. Protein Tyrosine Phosphatases: The Problems of a Growing Family. 1-87-93 Cold Spring Harbor Laboratory Press, 1992.
77. Riggleman R, Schedl P, Wieschaus E. Spatial expression of the Drosophila segment polarity gene *armadillo* is post-transcriptionally regulated by *wingless*. Cell 1990; 63:549-560.
78. Ingham P. Segment polarity genes and cell patterning within the *Drosophila* body segment. Curr Biol 1991; 1:261-267.
79. Nusse R, Varmus H. Wnt genes. Cell 1992; 69:1073-1087.
80. Siegfried E, Wilder EL, Perrimon N. Components of wingless signalling in Drosophila. Nature 1994; 367:76-79.
81. Peifer M, Sweeton D, Casey M, Wieschaus E. Wingless signal and Zeste-white 3 kinase trigger opposing changes in the intracellular distribution of armadillo. Development 1994b; In press.
82. Noordermeer J, Klingensmith J, Perrimon N, Nusse R. dishevelled and armadillo act in the Wingless signalling pathway in *Drosophila*. Nature 1994; 367:80-83.
83. Peifer M. The product of the Drosophila segment polarity gene armadillo is part of a multi-protein complex resembling the vertebrate adherens junction. J Cell Sci 1993.
84. Oda H, Uemura K, Shiomi K, Nagafuchi A, Tsukita S, Takeichi M. Identification of a Drosophila homolog of alpha-catenin and its association with the armadillo protein. J Cell Biol 1993; 121:1133-1141.
85. Cowin P, Mechanic S. In: Citi S, ed. Molecular Mechanism of Epithelial Cell Junctions: From Development to Disease. Austin: RG Landes, 1994.
86. Olson DJ, Christian JL, Moon RT. Effect of Wnt-1 and related proteins on gap junctional communication in *xenopus* embryos. Science 1993; 252:1173-1176.
87. Williams EJ, Walsh F, Doherty P. Tyrosine kinase inhibitors can differentially inhibit integrin-dependent and CAM-stimulated neurite outgrowth. J Cell Biol 1994; 124:1029-1037.
88. Saffell JL, Walsh F, Doherty P. Direct Activation of second messenger Pathways Mimics cell adhesion molecule-dependent Neurite Outgrowth. J Cell Biol 1992; 118:663-670.
89. McCrea PD, Brieher WM, Gumbiner BM. Induction of a secondary body axis in Xenopus by antibodies to β-catenin. J Cell

Biol 1993; 123:477-484.

90. McMahon AP, Moon RT. Ectopic expression of the proto-oncogene int-1 in xenopus embryos leads to duplication of the embryonic axis. Cell 1989; 58:1075-184.

91. Bradley RS, Cowin P, Brown AMC. Expression of Wnt-1 in PC12 cells results in modulation of plakoglobin and E-cadherin and increased cellular adhesion. J Cell Biol 1993; 123:1857-1865.

92. Brown AMC, Wildon RS, Prendergast TJ, Varmus HE. A retrovirus vector expressing the putative mammary oncogene int-1 causes partial transformation of a mammary epithelial cell line. 1986; 46:1001-1009.

93. Su LK, Vogelstein B, Kinzler KW. Association of the APC tumor suppressor protein with catenins. Science 1993; 262: 1734-1737.

94. Rubinfeld B, Souza B, Albert I, Muller O, Chamberlain S, Masiarz FR, Munemitsu S, Polakis P. Association of the APC gene product with beta-catenin. Science 1993; 262:1731-1734.

95. Kinsler KW, Nilbert MC, Su L, Vogelstein B, Bryan TM, Levy DB, Smith KJ, Presinger AC, Hedge P, McKechnie D, Finniear R, Markham A, Groffen J, Boguski MS, Altshul SF, Hirii A et al. Identification of FAP locus genes from chromosome 5q21. Science 1991; 253:661-669.

96. Nishisho I, Nakamura Y, Miyoshi Y, Miki Y, Ando H, Hirii A, Koyama K, Utsunomiya J, Baba S, Hedge P, Markham A, Krush A, Peterson G, Hamilton S , Nilbert MC, Levy DB et al. Mutations of chromosome 5q21 genes in FAP and colorectal cancer patients. Science 1991; 253:665-669.

CHAPTER 5

Hemidesmosomes: Form, Function and Assembly

Mikayo Langhofer, Scott E. Baker,
Susan B. Hopkinson, Jonathan C.R. Jones

INTRODUCTION

Hemidesmosomes are cell-extracellular matrix adhesion structures located on the basal surface of epithelial cells. They are present in stratified squamous, pseudostratified, transitional epithelia and some glandular tissue, and are especially abundant in tissues subject to mechanical stress like the epidermis and cornea. Hemidesmosomes were first called bobbins because, in the electron microscope, they looked like thread holders from sewing machines.[1] Subsequently, they were named hemidesmosomes due to their morphologic similarity to desmosomes (cell-cell adhesion structures; chapters 1 and 3). This similarity, however is primarily superficial, as it has been demonstrated immunochemically that (with one or possibly two exceptions, which will be discussed below) the major components of hemidesmosomes and desmosomes are quite distinct.[2]

When viewed by electron microscopy, the hemidesmosome is composed of a tripartite electron dense plaque next to the plasma membrane on the basal surface of the cell (Fig. 5.1). Intermediate filaments appear to insert into the cytoplasmic end of the plaque. In fact, one function of the hemidesmosome is to tether the cytoskeleton to the cell surface. Subjacent to the plaque extracellularly, anchoring filaments span the lamina lucida of the basement membrane zone (BMZ), linking the plasma membrane and the lamina densa. In the lamina lucida the so called subbasal dense plate, a thin linear density, parallels the plasma membrane of the hemidesmosome. Anchoring fibrils, composed of collagen type VII, are present in the matrix beneath the lamina densa.[3] Together the tripartite hemidesmosomal plaque and associated structures including intermediate filaments, anchoring filaments and anchoring fibrils is known as an adhesion complex.[4]

Molecular Biology of Desmosomes and Hemidesmosomes, edited by Jane E. Collins and David R. Garrod.

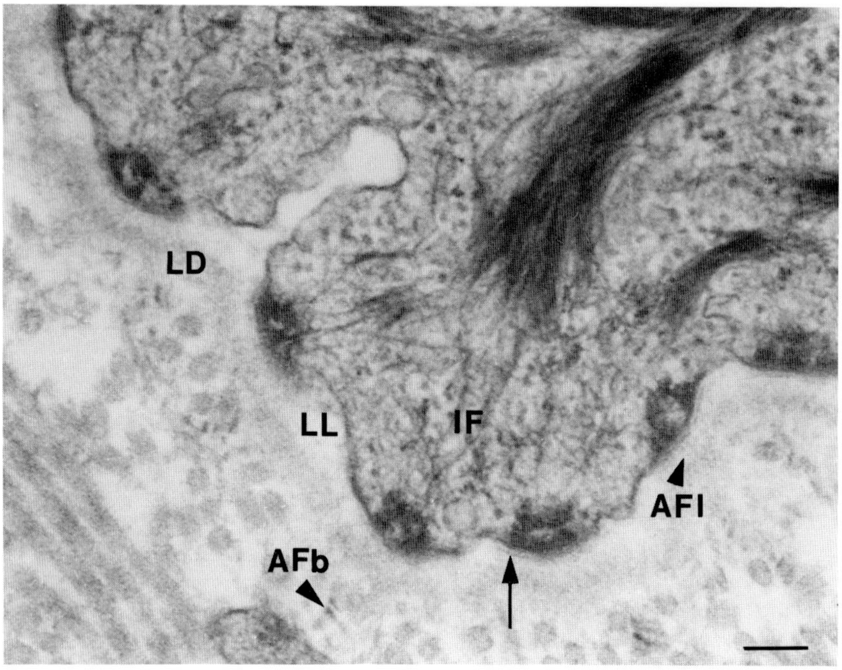

Fig. 5.1. Electron micrograph of hemidesmosomes in bovine tongue mucosa. Intermediate filaments (IF) are associated with the cytoplasmic aspect of the hemidesmosomes. The basement membrane zone is comprised of the lamina lucida (LL) and the lamina densa (LD). In the lamina lucida, subjacent to the hemidesmosomes are anchoring filaments (AFl, arrow head) and a sub-basal dense plate (arrow). In the dermal connective tissue are anchoring fibrils (AFb, arrow head). Bar, 100 nm.

Hemidesmosomes are of fundamental importance in cell-substrate adhesion. In disease states such as junctional epidermolysis bullosa (JEB) as well as the autoimmune diseases bullous pemphigoid (BP) and herpes gestationis (HG), disruption of hemidesmosomes compromises the attachment of the epidermis to the dermis.[5-13] Furthermore, during wound healing the assembly of the hemidesmosome is a necessary prerequisite to firm attachment of an epidermal sheet to the connective tissue.

Over the last few years a number of hemidesmosomal components have been identified at the cellular and molecular level (Fig. 5.2). These proteins will be described below. In addition we will discuss hemidesmosome morphogenesis during development and wound healing, as well as a cultured cell model for the study of hemidesmosomes in vitro.

HEMIDESMOSOME COMPONENTS

Bullous Pemphigoid (BP) Antigens

Nearly ten years ago, it was shown that autoantibodies in the serum of patients afflicted with bullous pemphigoid (BP) recognize hemidesmosomes.[14-16] These autoantibodies recognize two epidermal antigens of 230 kD (BPAG1 or BP230) and 180 kD (BPAG2 or BP180).[6,7,17,18]

BP230

BP230 was first characterized in the basement membrane of stratified squamous epithelia.[2,5,6] BP230 autoantibodies localize to the cytoplasmic portion of the hemidesmosome, close to the region where keratin-containing intermediate filament bundles appear to insert into the hemidesmosomal plaque.[18,19] BP230 is not exclu-

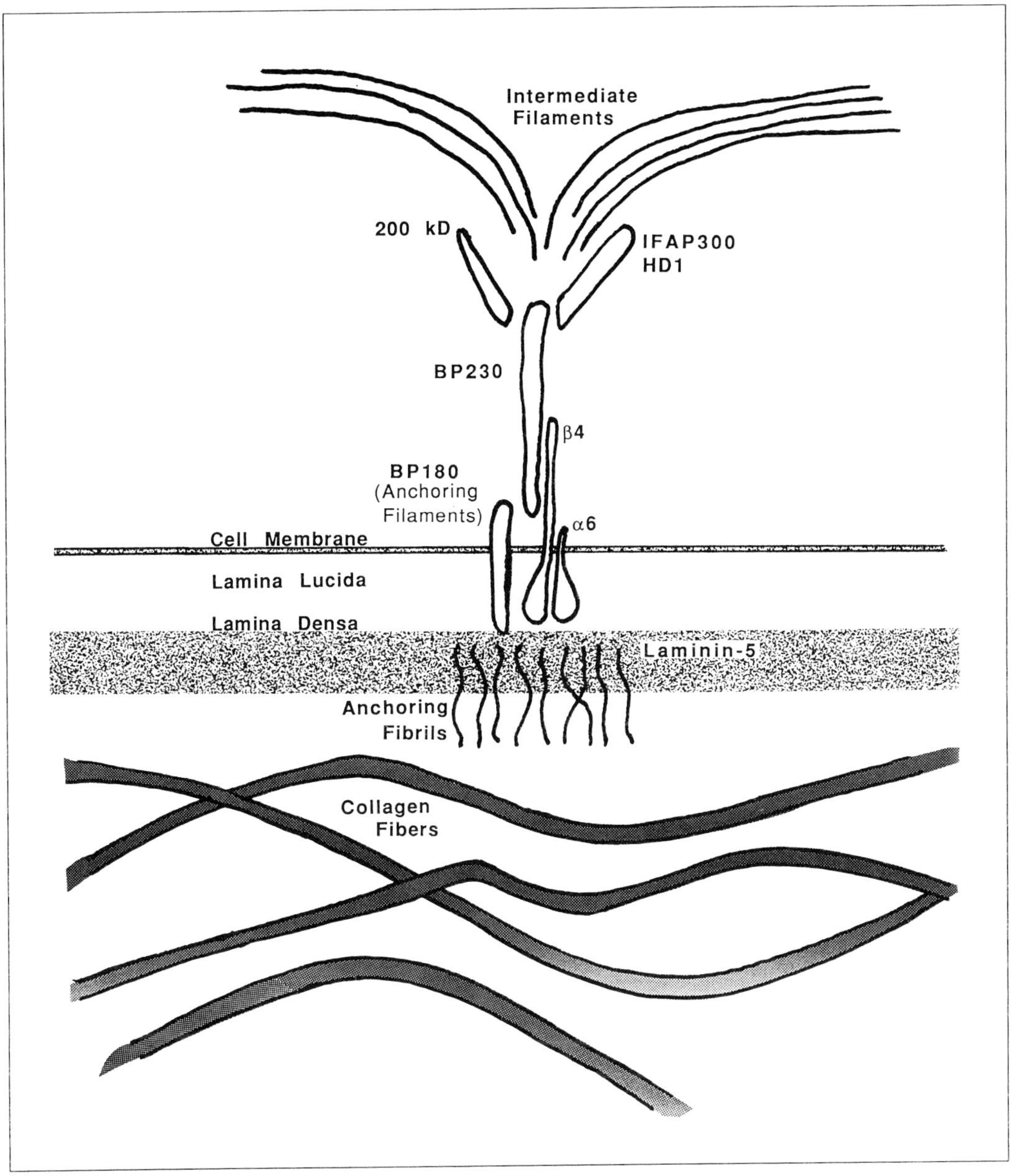

Fig. 5.2. A diagram of the relative location of the major polypeptides in the hemidesmosome. Some liberties have been taken both with putative protein-protein interactions, as well as exact protein location. In addition, IFAP300 and HD1 may be closely related proteins, and for this reason are indicated as one molecule.

sively expressed by epidermal cells but is also present in all epithelial cells that possess bona fide hemidesmosomes, including those of the cornea, bladder, trachea and breast.[20]

The entire sequence of BP230 has been determined,[21-25] and secondary structural analyses of this sequence predict that BP230 is a large molecule containing an α helical coiled-coil rod.[24,26] This is borne out by rotary shadow images of chromatographically purified BP230.[20] The carboxy terminus of BP230 contains two regions (termed B and C) of remarkable sequence and structural homology (40% sequence identity) to certain carboxy-terminal domains of desmoplakin I, a component of the plaque of the desmosome.[26] These repeats have the same regular linear distributions of acidic and basic amino acids found in the 1B segment of keratin intermediate filament subunits. Plectin, an intermediate filament associated protein, also shares sequence homologies with the B and C domains of BP230.[27] Indeed, it has been hypothesized that all three proteins interact with intermediate filaments via such domains.[19,28] This is certainly consistent with the localization of BP230 in the hemidesmosome plaque.[18] However, to date, of the three, only plectin has been shown to bind intermediate filaments in vitro.[29]

Recently, evidence for the existence of BP230 isoforms has been presented.[30] In addition to the BP230 message, Hopkinson and Jones[30] have identified an mRNA product of the single gene encoding BP230 which lacks the last 1800 bases of the BP230 message including the sequence encoding the B and C regions of BP230 that show homology to comparable domains in desmoplakin I and plectin. Intriguingly, the protein product of this second BP230 message appears to colocalize with keratin containing intermediate filaments in a pancreatic carcinoma epithelial cell line called FG, suggesting that the B and C domains of BP230 may not be necessary for interaction with intermediate filaments.

BP180

The other major epithelial protein recognized by BP autoantibodies is BP180. Sequence analysis suggests that BP180 is a type II transmembrane protein; the carboxy-terminus of the protein is extracellular while its amino-terminus is intracellular.[31-36] Proteolysis studies, cytoplasmic localization of N-terminal antibodies and extracellular localization of C-terminal antibodies by immunoelectron microscopy support the sequence analysis.[33,37] The extracellular domain of BP180 contains collagen-like repeats of variable length, which has led to BP180 being designated type XVII collagen.[34,36] It would appear likely that these collagenous extracellular domains are sites of interaction with other basement membrane zone components such as the collagens and laminins.[33,34]

In the cytoplasmic region of BP180, close to its predicted membrane spanning domain, four cysteine residues are ideally placed to facilitate either homophilic and/or heterophilic interactions.[33] Such interactions would allow BP180 to form macromolecular complexes; in fact, this has led to speculation that the collagen domains of BP180 assemble into anchoring filaments of the lamina lucida.[38]

A recent study by Liu et al[10] has shed light on the role that BP180 autoantibodies may play in pathogenesis of the disease, and in so doing, they have also identified a functionally important domain within the BP180 molecule. The majority of BP sera contain autoantibodies which recognize a non-collagenous extracellular domain of BP180 that lies close to the membrane of the hemidesmosome. When antibodies against a peptide contained in this domain are injected into mice, the mice develop blisters which resemble those present in BP patients, implicating such autoantibodies in the development of BP blisters.[10]

A 125kD Anchoring Filament Associated Protein

A hemidesmosomal polypeptide of 125kD was first identified using a monoclonal antibody prepared from mice immunized with a hemidesmosome enriched preparation derived from bovine tongue.[18] Immunoelectron microscopy places this polypeptide extracellular to the hemidesmosomal plaque, near the plasma membrane, suggesting it may be a structural or anchoring filament-associated protein.[39] In organ culture, anti-125 kD antibodies cause complete detachment of epithelial cells from the dermis, as well as blockage of hemidesmosome formation suggesting that the 125kD protein is a cell adhesion molecule.[39] Some investigators have even proposed that the 125kD molecule is the ligand for the integrin $\alpha6\beta4$ (see below for a description of other candidates for the $\alpha6\beta4$ ligand).[39]

A 200 kD Plaque Component, IFAP300, Plectin and HD1

A 200 kD polypeptide is associated with the cytoplasmic portion of the hemidesmosomal plaque.[40] This polypeptide is immunologically distinct from hemidesmosomal components recognized by BP autoantibodies.[40] Its function in the hemidesmosome has not been determined.

A number of molecules, including BP230, have been proposed as intermediate filament-hemidesmosomal plaque linkers. The best candidate to date is a 300 kD intermediate filament binding protein termed IFAP300.[41] Initially characterized in fibroblasts,[42] IFAP300 antibodies localize to hemidesmosomes at the ultrastructural level at sites where intermediate filaments appear to attach to the hemidesmosomal plaque (Fig. 5.3).[41] Moreover, purified IFAP300 sediments with assembled keratins under conditions in which IFAP300 alone does not pellet, indicating that IFAP300 binds intermediate filaments.[41]

A partial sequence of IFAP300 has revealed that IFAP300 is related to but not the same as another intermediate filament binding protein called plectin.[41] Antibodies against plectin have also been localized to the site of hemidesmosomes.[43] Thus it is possible that multiple intermediate filament associated proteins are involved in intermediate filament-plaque interaction.

Another potential intermediate filament-hemidesmosomal plaque linker protein is a 500 kD protein termed HD1.[44] This protein is enriched in bovine corneal epithelial cell hemidesmosomes. Like IFAP300 and BP230, HD1 localizes to the cytoplasmic side of the hemidesmosome plaque where intermediate filaments insert. HD1, like IFAP300, is not exclusive to cells which contain hemidesmosomes as defined ultrastructurally; for example, it is

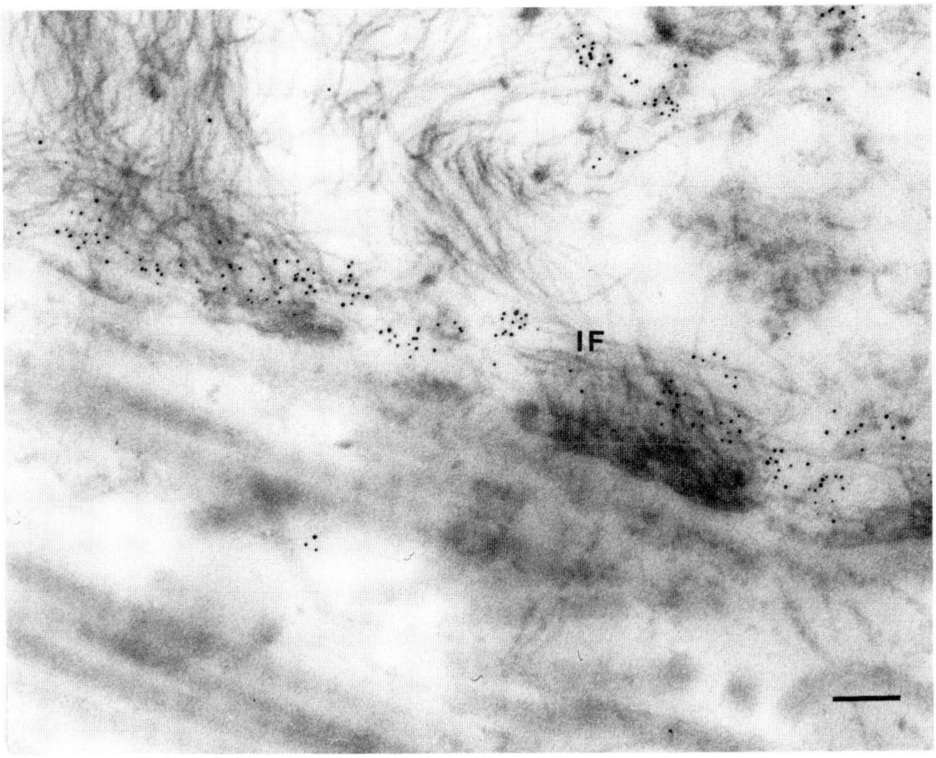

Fig. 5.3. Immunogold localization of IFAP300 in bovine tongue mucosa reveals gold particles are concentrated at the plaque region into which intermediate filaments interact. IF, intermediate filaments. Bar, 200 nm.

found in endothelial and glial cells in regions of cell-matrix interaction. Based on this finding, Hieda et al[44] have proposed the existence of a second, or type II hemidesmosome that is BP antigen negative but HD1 positive.

INTEGRINS

Integrins are a family of integral membrane glycoproteins mediating cell-cell and cell-extracellular matrix adhesion. Each integrin is a heterodimer comprised of an α and a β subunit, whose various combinations in different tissues give rise to receptors with specific binding capacities.[45-48] The ligands recognized by these heterodimer receptors are extracellular matrix components.

The integrin α6β4 was initially observed to be preferentially expressed in epithelial tissues.[49-53] Indirect immunofluorescence of stratified squamous epithelia with anti-α6β4 antibodies reveal an intense fluorescence at the epithelial connective tissue junction.[50,51,54] The pattern is very similar to that generated by antibodies directed against hemidesmosomal plaque elements. Using immunoelectron microscopy several groups have shown that the integrin α6β4 is concentrated in the hemidesmosome (Fig. 5.4).[55-57]

804G CELL MATRIX/KALININ/NICEIN/LAMININ ISOFORMS

The nature of the ligand for the integrin α6β4 is still unclear. In at least one epithelial cell type, α6β4 appears capable of binding EHS tumor type laminin[58] (recently renamed laminin-1).[59] However, this does not appear to be the "normal" α6β4 ligand. Rather, it is likely that a laminin isoform which is associated with hemidesmosomes is the α6β4 ligand. This isoform has been identified variously as nicein, kalinin, epiligrin and "804G matrix" (recently renamed laminin-5).[59-63] The term 804 cell matrix refers to the extracellular matrix produced by the rat bladder cell line 804G, which readily assembles hemidesmosomes in culture (Fig. 5.5).[63] Laminin-5 is a heterotrimer composed of novel A, B1 and B2 laminin chains with a non-reduced molecular weight of over 450 kD.[60-63] In human keratinocytes, laminin-5 is apparently synthesized in three high molecular weight complexes of 460, 440 and 400 kD.[61,64] The cell-associated form is 460 kD and is composed of nonidentical subunits of 200,

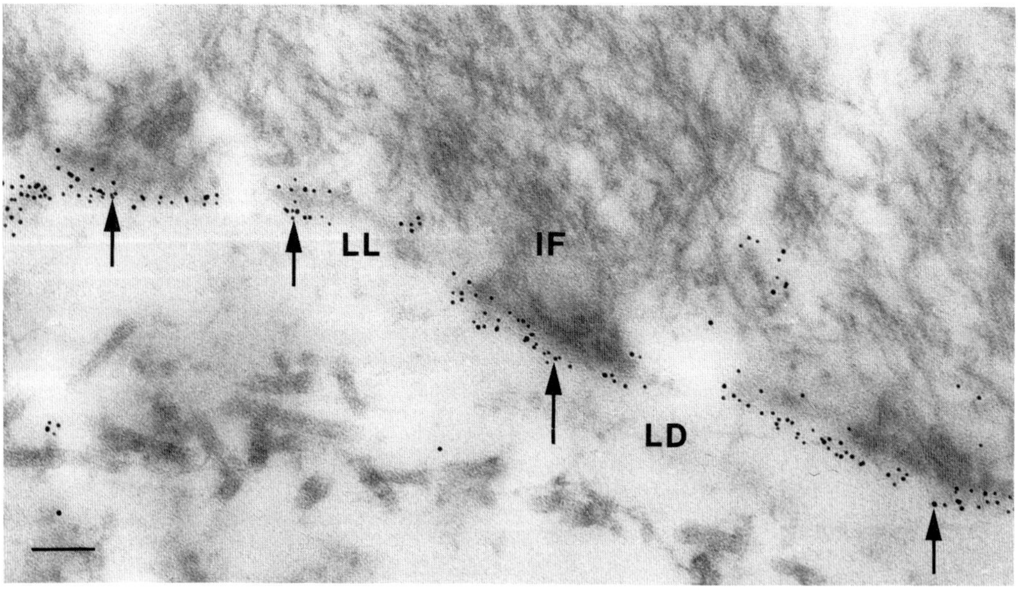

Fig. 5.4. Immunogold localization of integrin β4 in bovine tongue mucosa. Gold particles (arrows) are present underlying the hemidesmosomes, in the lamina lucida (LL). LD, lamina densa. Bar, 100 nm.

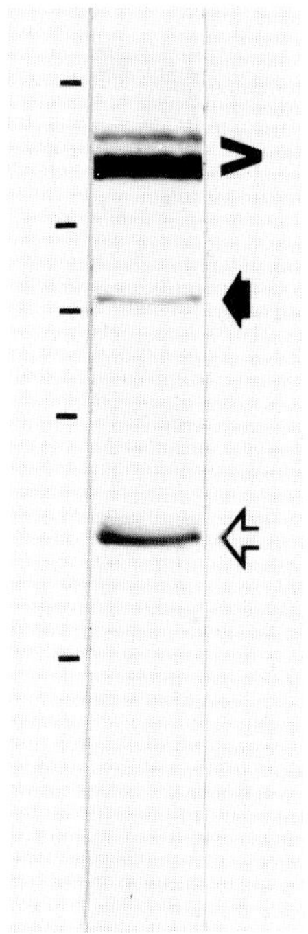

Fig. 5.5. A preparation of laminin-5 rich 804G cell matrix was prepared according to the method of Gospodarowicz,[126] processed for SDS-PAGE, and silver stained. The major polypeptides in the matrix preparation are 150, 140, 135, 100 and 55 kD. Rat laminin-5 is comprised of the three bands of 150, 140 and 135 kD The 100 kD chain (arrow) is a proteolytic fragment of the 140 kD band.[63] The 55 kD polypeptide (open arrow) has not yet been characterized. Molecular mass standards of 200, 116, 97, 66 and 45 kD are indicated.

155 and 140 kD linked by interchain disulfide bonds and is secreted by keratinocytes onto the culture substratum. The 440 kD form is present in the conditioned medium of cultured keratinocytes, and is derived by extracellular processing of the 200 kD subunit to 165 kD, a step which also happens in skin organ culture. The 400 kD form is most closely related to that purified from tissue, and is derived from the 440 kD form by extracellular processing of the 155 kD subunit to 105 kD.[61,64]

Sequence analysis of the human 155 and 105 kD laminin-5 chains (in rat these proteins migrate at 140 and 100 kD) has shown them to be truncated laminin-1 B2 chains (termed B2t).[63,65,66] The human 140 kD glycoprotein component of laminin-5 is a B1 chain,[67] while the 165 kD human protein (migrating at 155 kD in rat) is an A-chain variant showing approximately 22% identity to other A-chains.[68]

The evidence that laminin-5 is the ligand for $\alpha6\beta4$ integrin is circumstantial: laminin-5, BP antigens and the integrin heterodimer $\alpha6\beta4$ colocalize in both cultured human epithelial cells in "stable anchoring contacts", in hemidesmosomes in rat 804G cells (Fig. 5.6), and in the hemidesmosomes of tissue.[62,63,69,70] A number of laboratories have data from adhesion studies that indicate that the ligand for $\alpha6\beta4$ is laminin-5.[71,72] More importantly, the laminin-5 isoform-rich matrix, produced by 804G cells, induces the formation of hemidesmosomes in human epithelial cell lines in an $\alpha6\beta4$ dependent manner (see below, unpublished data). These results conflict with recent reports which have concluded that laminin-5 functions as the ligand for $\alpha_3\beta_1$ and $\alpha_6\beta_1$ integrin but not $\alpha6\beta4$.[62,69,73,74] It will be necessary to assess ligand binding at the biochemical level to clarify these issues.

HEMIDESMOSOME MORPHOGENESIS AND DYNAMICS IN DEVELOPMENT, WOUND HEALING AND CELL CULTURE

Much effort has gone into the study of hemidesmosome assembly during skin development and wound healing. In developing human fetal skin, the lamina lucida, lamina densa and anchoring filaments are present at 6 weeks gestational age.[75] Laminin-1, type IV collagen and fibronectin are also present at this early time point. BP antigens (as detected by immunofluorescence using whole BP sera), hemidesmosomes and anchoring fibrils do not appear until weeks 10-11 when the epidermis is stratifying from a 2 to a 3 layered epithelium.[75] In developing fetal rabbit cornea, hemidesmosomes and anchoring fibrils also appear synchronously;[76] at the immunochemical level, α_6 and BP antigens appear synchronously and colocalize with type VII collagen.[55]

Various models of wound healing have been used to study hemidesmosome assembly. These include suction induced blister, epithelial graft (where epithelial sheets are placed on connective tissue), incision and epiboly (the process of epithelial outgrowth in skin or corneal cultures) models. In some of these models such as suction induced blisters, many of the basement membrane zone components remain, frequently appearing to

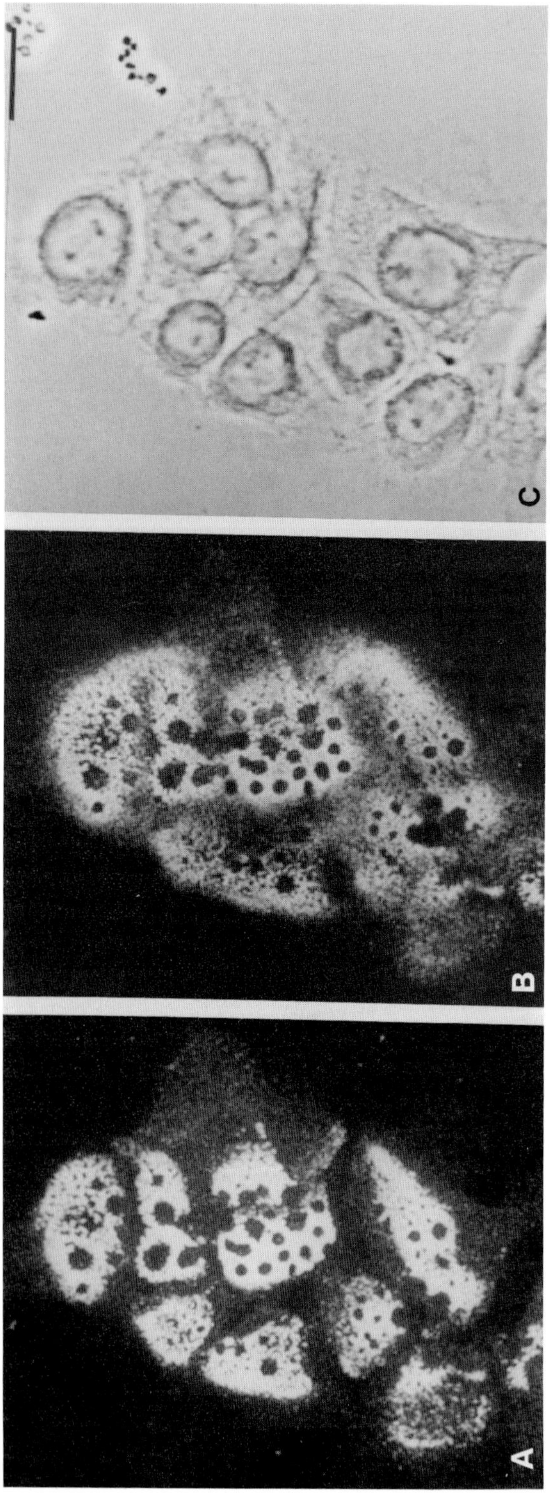

Fig. 5.6. 804G cells, derived from rat bladder carcinoma, were processed for double label indirect immunofluorescence using antibodies against BP230 (A) and an antiserum that recognizes 804G cell matrix (B). At the basal attached surface of the 804G cells, the anti-BP230 antibodies (A) generate a dramatic staining pattern that is coincident with that generated by the anti-matrix antibodies (B). Bar in C, 10 µm.

act as nucleation sites for hemidesmosomes.[77] In others, such as epiboly models, a new basement membrane has to be formed in addition to hemidesmosome assembly during the process of epithelialization of "wounded" surfaces.

Hemidesmosome Assembly in Suction Induced Blisters and Following Epithelial Grafting onto Basement Membranes

During the epithelialization of suction induced blisters, hemidesmosomes assemble rapidly within 1-2 days.[78,79] Hemidesmosome formation is preceded by the appearance of anchoring filaments. Likewise, in rabbit corneal epithelial sheets placed on denuded corneal stroma, hemidesmosomes form rapidly (within 6 hours in culture).[77] In this corneal model, by 24 hours the number of hemidesmosomes is 95% that of control cornea. Moreover, between 80 and 86% of the hemidesmosomes form directly over preexisting anchoring fibrils in the denuded corneal stroma. From these results Gipson et al[77] have hypothesized that anchoring fibrils act as nucleation sites for new hemidesmosomes. The speed of assembly of hemidesmosomes observed suggests that there are either large pools of hemidesmosome elements or that hemidesmosome components may be reused in the grafted cells. There is some evidence for the latter possibility.[14,77,80-83] In order to prepare individual epithelial cells or cell sheets for grafting, the cells are generally detached from either connective tissue or culture plates using enzymes. Some workers have reported that hemidesmosomes remain along the cell surface or become endocytosed in the detached cells.[14,77,80-83] In either case it is possible that the residual hemidesmosomes could simply be used again upon the reattachment of the cell to a suitable substrate.

Hemidesmosome Assembly in Incision Wounds, "Epibolies," and Following Epithelial Grafting onto Connective Tissue Lacking Basement Membrane

It is becoming increasingly clear that hemidesmosome assembly is influenced by the substratum upon which an epithelial cell sits. For example, corneal epithelial cells from chick embryos cultured on lens capsule, gelatin (type I collagen), and bovine corneal extracellular matrix form only "immature" hemidesmosomes.[84] In a study by Gipson et al,[77] viable sheets of rabbit corneal epithelial cells cultured on Descemet's membrane and lens capsule for 24 hours fail to form hemidesmosomes. Since neither Descemet's membrane, or lens capsule, contain anchoring fibrils, these results have been taken as further evidence that anchoring fibrils act as nucleation sites for the assembly of new hemidesmosomes.[77]

In vivo the lack of an intact basement membrane retards, although does not block, hemidesmosome assembly in epithelial cells populating wound sites. For example, epidermal cells, placed on wound beds where the connective tissue has been excised to the muscle facia, assemble hemidesmosomes by about 6 days after grafting compared with the 1-2 days in suction blisters.[78,79,85] Electron microscopic studies of the forming epidermal-dermal junction, both in vitro and in vivo, have shown that hemidesmosomes appear prior to the lamina densa region of the basement membrane.[39,80,86-88] Subsequently, lamina densa material develops subjacent to the hemidesmosomal plaque. However, in certain instances of wound healing in the skin anchoring fibrils may take a year to return to the diameters and densities of those seen in normal skin.[85]

The order of appearance of the hemidesmosome and basement membrane zone antigens, as assessed by indirect immunofluorescence microscopy, during the healing of wounds, is controversial and different investigators have obtained different results. However, in general, laminin-1, collagen type IV, and collagen type VII appear either synchronously with, or soon after, the appearance of BP antigens.[17,39,87-92] In this regard, the disparity of the published studies may well reflect differences in the sensitivities of antibodies, the availability of antibody epitope, and potential variability in hemidesmosome assembly among species (mouse, pig, rabbit, bovine, or human), or, between the different tissue examined (cornea versus skin).

The use of three dimensional cell culture systems have provided additional information on the genesis of the dermis and basement membrane zone.[93,94] In this model epidermally derived fibroblasts are grown on a nylon mesh in the presence of ascorbate, creating a collagen rich matrix resembling dermis in 3-4 weeks.[94] Type I collagen is present in the fibrils of diameters less than 34 nm while type II collagen was present in the fibers of 35-54 nm. Keratinocytes are seeded onto this fibroblast created dermis, and allowed to grow for 1 week submerged and at the air-liqiuid interface for an additional 4 weeks. Electron microscopic analysis of this in vitro derived skin reveals hemidesmosomes, anchoring filaments, a well developed lamina densa, and anchoring fibrils.[93] These data are consistent with the findings of Regauer et al[95] who demonstrated that anchoring fibrils are primarily a product of epidermal keratinocytes. However, there is still some debate concerning the cellular source of anchoring fibrils. Type VII procollagen, the precursor molecule of anchoring fibrils, is secreted by both keratinocytes and dermal fibroblasts in culture.[3] It is likely that the dermal fibroblasts and epidermal keratinocytes regulate each other in the formation of hemidesmosomes and the basement membrane zone, and that anchoring fibrils are contributed by both cell populations.

Cell Culture Models of Hemidesmosome Assembly and Dynamics

Most cells when grown in culture do not readily form hemidesmosomes, even though the protein components are present, as determined by immunofluorescence.[18,63] However, a few cell lines have been described which assemble hemidesmosomes in culture.[72,96] For example, 804G cells, a rat bladder carcinoma derived cell line, readily form hemidesmosomes when maintained on uncoated substrates in vitro.[96] The hemidesmosomes made by the 804G cells are morphologically and immunochemically similar to those seen in bladder cells in situ (Fig. 5.7).[96] Consequently, it has been possible to analyze the formation of hemidesmosomes in 804G cells at the morphological level under various cell culture conditions.

The fate of BP230 and BP180 in 804Gs has been examined, following closure of in vitro wounds, during mitosis, and during drug induced disruption of the cytoskeleton.[97] During in vitro wound healing, formed hemidesmosomes are lost as cells enter the wound site. However, once in the wound site, cells appear to stop migrating and form hemidesmosomes. First, hemidesmosomes reappear on the basal surface of the cells at their periphery, and later these hemidesmosomes "move" under the nucleus. During mitosis, 804G cells retain their hemidesmosomes, even though they appear to partially round up prior to cell division.[97] The same is true for epidermal cells in vivo.[98]

Intriguingly, when the cytoskeleton of 804G cells is disrupted with cytochalasin D, BP antigens and formed hemidesmosomes redistribute and concentrate at the cell periphery along fine cell extensions. Under these circumstances, they are still associated with keratin bundles, which extend along the cell projections. Riddelle et al[97] have speculated that hemidesmosomal plaques are mobile within the plasma membrane. They hypothesize that depolymerization of the microfilaments by cytochalasin D causes the keratin bundles to uncoil and extend into the cell projections. The hemidesmosomal plaques, attached to intermediate filaments would follow the keratin filaments. Obviously this can only occur if hemidesmosomal plaques are dynamic and can move within the plane of the cell membrane.

The role of $\alpha6\beta4$ in the formation of hemidesmosomes has been explored in 804G cells.[56] Antibodies directed against the extracellular domains of $\alpha6\beta4$ block assembly of hemidesmosomes in 804G cells suggesting that $\alpha6\beta4$ is involved in nucleating hemidesmosome formation.[56] This is supported by antibody blocking studies in a bovine corneal epiboly wound model; when polyclonal anti-β_4 antibodies are included in the culture medium, the migration of epithelial cells is undisturbed, but the formation of hemidesmosomes and the expression of collagen type VII is inhibited.[91] The anti-β_4 antibodies also disrupt established hemidesmosomes in nonmigrating cells in the unwounded area of the explant.

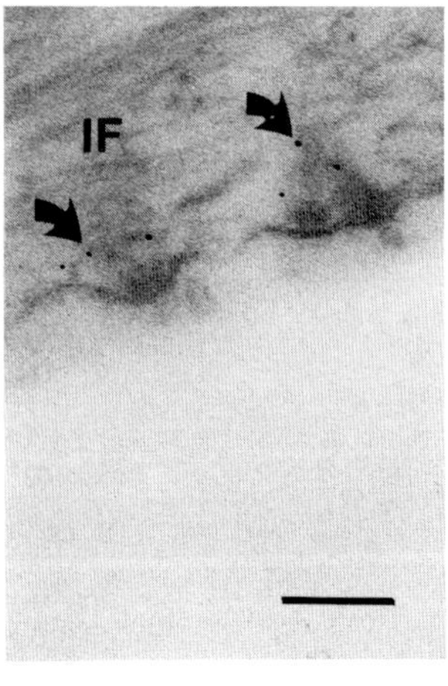

Fig. 5.7. 804G cells were processed for immunogold electron microscopy localization using a monoclonal antibody against BP230, called 5E.[22,127] The gold particles (arrows) are located on the cytoplasmic aspect of hemidesmosome-like structures that occur along regions of cell-substrate interaction. IF, intermediate filaments. Bar, 200 nm.

Furthermore, the inclusion of monoclonal anti-α_6 antibodies in the culture media causes epithelial cells to detach from the connective tissue.[91]

Langhofer et al[63] have examined the role of a laminin-5 rich matrix derived from 804G cells in hemidesmosome formation. $\alpha6\beta4$ and the BP antigens of human epidermal cells plated onto 804G matrix redistribute in a pattern identical to the distinctively patterned matrix present below the cells.[63] Moreover, the 804G cell matrix induces the formation of "mature" hemidesmosomes in human epithelial cells, which normally only form "prehemidesmosomes" when cultured on other matrices such as collagen type I or IV, laminin-1, or fibronectin.[63] In addition, when human epithelial cells are grown on the laminin-5 rich cell matrix in the presence of anti-α_6 monoclonal antibodies, BP antigens and $\alpha6\beta4$ no longer colocalize with the 804G cell matrix (unpublished data), indicating that the formation of hemidesmosomes is $\alpha6\beta4$ dependent.

804G cells have been used to delineate functional domains of various hemidesmosomal components. Full length and truncated human β_4 cDNA have been expressed in 804G cells in order to examine the function of the cytoplasmic and extracellular domains of the molecule.[99] Recombinant β_4 lacking almost all of its cytoplasmic domain associates with α_6, and reaches the cell surface, but remains diffusely distributed. β_4 missing almost all of the extracellular portion does not associate with α_6, but is correctly targeted to the hemidesmosome.[99] Further truncation studies demonstrate that a 303-amino acid region in the cytoplasmic domain is necessary for $\alpha6\beta4$ assembly into the hemidesmosome.[99]

HEMIDESMOSOMES, THE INTEGRIN $\alpha6\beta4$ AND LAMININ-5 IN DISEASE

Autoimmune Blistering Diseases

The autoimmune blistering diseases bullous pemphigoid (BP) and herpes gestationis (HG), described earlier, are characterized by subepidermal blisters, which show a clean separation of the epidermis from the dermis through the lamina lucida. Autoantibodies present in the sera of BP and HG patients recognize BP180 and BP230 as we have already discussed. However, the role of these autoantibodies in the pathogenesis of the diseases is controversial. Certainly, it is difficult to imagine that antibodies against BP230, a cytoplasmic protein, could be causative in BP. BP180 autoantibodies are more likely to be pathogenic, evidence for which Liu et al[10] have provided (see above).

Epidermolysis bullosa acquisita (EBA) is another autoimmune blistering disease, very similar in appearance to dystrophic epidermolysis bullosa (discussed below), characterized by blisters with a cleavage plane in the upper dermis. Autoantibodies present in the sera of these patients recognize the NC-1 domain of the type VII collagen molecule, and are believed to be pathogenic.[100-104]

Genetic Blistering Diseases

Disorders of the skin characterized by blistering following mechanical trauma have been classified together as epidermolysis bullosa (EB).[105] Except for EBA, discussed above, these diseases have a genetic basis and are subclassified based upon the site of the blister cleavage plane in the skin. Skin splits within the basal cells of the epidermis are classified as EB simplex. Patients suffering from this disorder have been shown to have mutations in keratin.[106,107] When the split occurs in the basement membrane zone of the lamina lucida (Fig. 5.8), it is classified as junctional EB (JEB). Immunofluorescence analysis using skin sections from patients diagnosed with JEB reveal an absence of laminin-5, and furthermore, the expression of $\alpha6\beta4$ and BP antigens is discontinuous.[9] More recent studies have identified the defect in Herlitz JEB, a subgroup of JEB, to be a consequence of altered levels of mRNA encoding for one of the laminin-5 chains. Moreover, in one family, a point

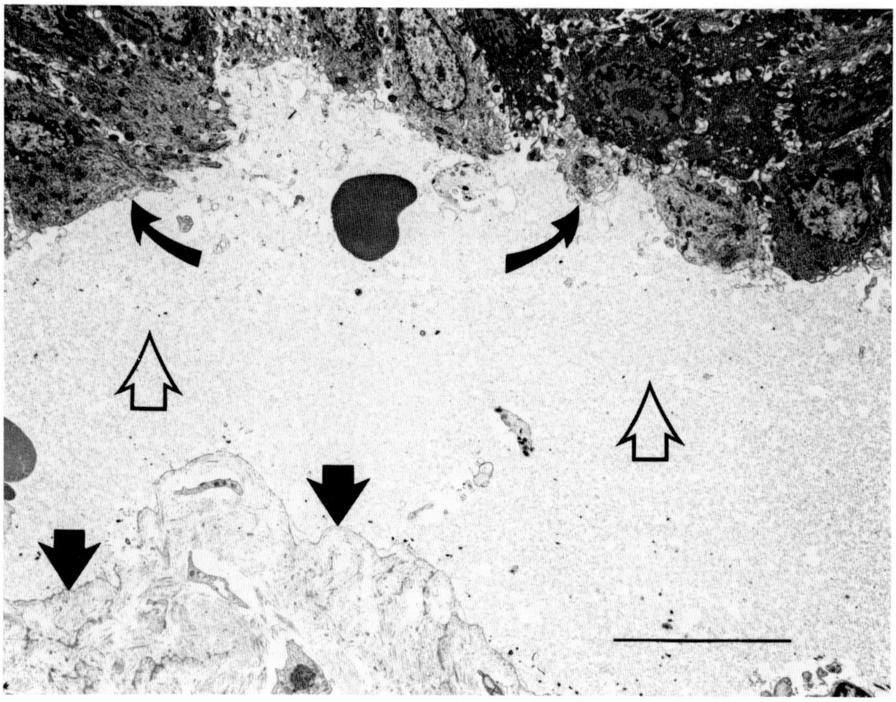

Fig. 5.8. Electron micrograph of a junctional epidermolysis bullosa blister (JEB). The skin has split at the basement membrane zone of the lamina lucida. The epidermis (curved arrows) is detached from the dermal connective tissue while the lamina densa appears intact (solid arrows). Open arrows, blister cavity. Bar, 10 μm.

mutation in exon 3 produces a stop codon in the B2 chain of laminin-5, preventing full length chain expression.[11,12,108]

In dystrophic EB, the cleavage plane is in the upper dermis. Patients with this disease often show a reduction in, or a complete lack of anchoring fibrils.[109] Linkage studies have demonstrated that the genetic defect responsible for dystrophic EB is in the collagen type VII gene.[110-114]

FUTURE DIRECTIONS

The regulation of expression of BP antigens is now under study. For example, TGF-β upregulates expression of both BP230 and BP180.[115] Furthermore, a keratinocyte protein that binds a regulatory element of the BP230 gene has been identified.[116] Intriguingly, the identification of a second protein product of the single gene encoding BP230 in a tumor cell line has led to the speculation that differential expression of the two gene products impacts on the ability of a cell to assemble hemidesmosomes.[30] The latter could explain the loss of hemidesmosomes in certain types of tumors. In this vein, the fate of the integrin α6β4 has been studied in different cancers but results are not consistent from one type of cancer to another, and should be further investigated. For example, in head and neck tumors there is an apparent over expression of α6β4 while in certain skin cancers expression of the α6β4 heterodimer is lost.[117-123] To add to the confusion, Sager et al[124] identified α_6 as a tumor suppressor gene in breast cancer by using a differential expression cloning method, while others have suggested that enhanced expression of α_6 is linked to tumor progression.[123]

We now need to know more about the way proteins interact in the hemidesmosome. To date, IFAP300 has been shown to interact with intermediate filaments, but otherwise we have no evidence for any other protein-protein interaction in the hemidesmosome. Finally, we know that the function of the hemidesmosome is in adhesion, but like the focal contact, it is tempting to speculate that it may be involved in signaling from the epithelium to the connective tissue and visa versa.[125] If this is the case we expect increased interest in the hemidesmosome to be generated in the next few years.

Acknowledgments

The Jones laboratory is supported by grants from the National Institute of Health and the American Cancer Society. We are grateful to Dr. T. Hashimoto for 5E monoclonal antibody.

References

1. Weiss P, Ferris W. Electronmicrographs of larval amphibian epidermis. Exp Cell Res 1954; 6:546-549.
2. Jones JCR, Yokoo KM, Goldman RD. Is the hemidesmosome a half desmosome? An immunological comparison of mammalian desmosomes and hemidesmosomes. Cell Motil Cytoskel 1986; 6:560-569.
3. Sakai LY, Keene DR, Morris NP et al. Type VII collagen is a major structural component of anchoring fibrils. J Cell Biol 1986; 103:1577-1586.
4. Gipson IK, Spurr-Michaud S, Tisdale A et al. Reassembly of the anchoring structures of the corneal epithelium during wound repair in the rabbit. Invest Ophthalmol Vis Sci 1989; 30:425-434.
5. Stanley JR, Hawley-Nelson P, Yuspa SH et al. Characterization of bullous pemphigoid antigen: a unique basement membrane protein of stratified squamous epithelia. Cell 1981; 24:897-903.
6. Stanley JR, Woodley DT, Katz SI. Identification and partial characterization of pemphigoid antigen extracted from normal human skin. J Invest Dermatol 1984; 82:108-111.
7. Labib RS, Anhalt GJ, Patel HP et al. Molecular hetrogeneity of the bullous pemphigoid antigens as detected by immunoblotting. J Immunol 1986; 136:1231-1235.
8. Chapman SJ, Leigh IM, Tidman MJ et al. Abnormal expression of hemidesmosome-like structures by junctional epidermolysis bullosa keratinocytes in vitro. Br J Dermatol 1990; 123:137-144.
9. Jonkman MF, de Jong MCJM, Heeres K et al. Expression of integrin $\alpha6\beta4$ in junctional epidermolysis bullosa. J Invest Dermatol 1992; 99:489-496.
10. Liu Z, Diaz LA, Troy JL et al. A passive transfer model of the organ-specific autoimmune disease, bullous pemphigoid, using antibodies generated against the hemidesmosomal antigen, BP180. J Clin Invest 1993; 92:2480-2488.
11. Aberdam D, Galliano M-F, Vailly J et al. Herlitz's junctional epidermolysis bullosa is linked to mutations in the gene (LAMC2) for the $\gamma2$ subunit of nicein/kalinin (laminin-5). Nature Genet 1994; 6:299-304.
12. Baudoin C, Miquel C, Blanchet-Bardon C et al. Herlitz junctional epidermolysis bullosa keratinocytes display heterogeneous defects of nicein/kalinin gene expression. J Clin Invest 1994; 93:862-869.
13. Tidman MJ, Eady RAJ. Hemidesmosome heterogeneity in junctional epidermolysis bullosa revealed by morphometric analysis. J Invest Dermatol 1986; 86:51-56.
14. Mutasim DF, Takahashi Y, Labib RS et al. A pool of bullous pemphigoid antigen(s) is intracellular and associated with the basal cell cytoskeleton-hemidesmosome complex. J Invest Dermatol 1985; 84:47-53.
15. Regnier M, Vaigot P, Michel S et al. Localization of bullous pemphigoid antigen (BPA) in isolated human keratinocytes. J Invest Dermatol 1985; 85:187-190.
16. Westgate GE, Weaver AC, Couchman JR. Bullous pemphigoid antigen localization suggests an intracellular association with hemidesmosomes. J Invest Dermatol 1985; 84:218-224.
17. Stanley JR, Alvarez OM, Bere EW Jr. et al. Detection of basement membrane zone antigens during epidermal wound healing in pigs. J Invest Dermatol 1981; 77:240-243.
18. Klatte DH, Kurpakus MA, Grelling KA et al. Immunochemical characterization of three components of the hemidesmosome and their expression in cultured epithelial cells. J Cell Biol 1989; 109:3377-3390.

19. Jones JCR, Green KJ. Intermediate filament-plasma membrane interactions. Curr Opinions Cell Biol 1991; 3:127-132.
20. Klatte DH, Jones JCR. Purification of the 230-kD bullous pemphigoid antigen (BP230) from bovine tongue mucosa: structural analysis and assessment of BP230 tissue distribution using a new monoclonal antibody. J Invest Dermatol 1994; 102:39-44.
21. Stanley JR, Tanaka T, Mueller S et al. Isolation of complementary DNA for bullous pemphigoid antigen by use of patients' autoantibodies. J Clin Invest 1988; 82:1864-1870.
22. Amagai M, Hashimoto T, Tajima S et al. Partial cDNA cloning of the 230-kD mouse bullous pemphigoid antigen by use of a human monoclonal anti-basement membrane zone antibody. J Invest Dermatol 1990; 95:252-259.
23. Tanaka T, Korman NJ, Shimizu H et al. Production of rabbit antibodies against carboxy-terminal epitopes encoded by bullous pemphigoid cDNA. J Invest Dermatol 1990; 94:617-623.
24. Sawamura D, Li K, Chu, M-L et al. Human bullous pemphigoid antigen (BPAG1). J Biol Chem 1991; 266:17784-17790.
25. Sawamura D, Li K-H; Nomura K et al. Bullous pemphigoid antigen: cDNA cloning, cellular expression, and evidence for polymorphism of the human gene. J Invest Dermatol 1991; 96:908-915.
26. Tanaka T, Parry DAD, Klaus-Kovtun V et al. Comparison of molecularly cloned bullous pemphigoid antigen to desmoplakin I confirms that they define a new family of cell adhesion junction plaque proteins. J Biol Chem 1991; 266:12555-12559.
27. Wiche G, Becker B, Luber K et al. Cloning and sequencing of rat plectin indicates a 466-kD polypeptide chain with a three-domain structure based on a central alpha-helical coiled coil. J Cell Biol 1991; 114:83-99.
28. Green KJ, Parry DA, Steinert PM et al. Structure of the human desmoplakins. Implications for function in the desmosomal plaque. J Biol Chem 1990; 265:2603-2612.
29. Foisner R, Leichtfried FE, Herrmann H et al. Cytoskeleton-associated plectin: in situ localization, in vitro reconstitution, and binding to immobilized intermediate filament proteins. J Cell Biol 1988; 106:723-733.
30. Hopkinson SB, Jones JCR. Identification of a second protein product of the gene encoding a human epidermal autoantigen. Biochem J 1994; 300:851-857.
31. Diaz LA, Ratrie H III, Saunders WS et al. Isolation of a human epidermal cDNA corresponding to the 180-kD autoantigen recognized by bullous pemphigoid and herpes gestationis sera. J Clin Invest 1990; 86:1088-1094.
32. Giudice GJ, Squiquera HL, Elias PM et al. Identification of two collagen domains within the bullous pemphigoid autoantigen, BP180. J Clin Invest 1991; 87:734-738.
33. Hopkinson SB, Riddelle KS, Jones JCR. Cytoplasmic domain of the 180-kD bullous pemphigoid antigen, a hemidesmosomal component: molecular and cell biologic characterization. J Invest Dermatol 1992; 99:264-270.
34. Giudice GJ, Emery DJ, Diaz LA. Cloning and primary structural analysis of the bullous pemphigoid autoantigen BP180. J Invest Dermatol 1992; 99:243-250.
35. Li K, Guidice GJ, Tamai K et al. Cloning of partial cDNA from mouse 180-kDa bullous pemphigoid antigen (BPAG2), a highly conserved collagenous protein of the cutaneous basement membrane zone. J Invest Dermatol 1992; 99:258-263.
36. Li K, Tamai K, Tan EML et al. Cloning of type XVII collagen. J Biol Chem 1993; 268:8825-8834.
37. Giudice GJ, Emery DJ, Zelickson BD et al. Bullous pemphigoid and herpes gestationis autoantibodies recognize a common non-collagenous site on the BP180 ectodomain. J Immunol 1993; 151:5742-5750.
38. Jones JCR, Asmuth J, Baker SE et al. Hemidesmosomes: extracellular matrix/intermediate filament connectors. Exp Cell Res 1994; 213:1-11.
39. Kurpakus MA, Stock EL, Jones JCR. Analysis of wound healing in an in vitro model: early appearance of laminin and a $125 x 10^3$ M_r polypeptide during adhesion complex formation. J Cell Sci 1990; 96:651-660.
40. Kurpakus MA, Jones JCR. A novel hemi-

desmosomal plaque component: tissue distribution and incorporation into assembling hemidesmosomes in an in vitro model. Exp Cell Res 1991; 194:139-146.
41. Skalli O, Jones JCR, Gagescu R et al. IFAP 300 is common to desmosomes and hemidesmosomes and is a possible linker of intermediate filaments to these junctions. J Cell Biol 1994; 125:159-170.
42. Yang, H-Y; Lieska N, Goldman AE et al. A 300,000-mol-wt intermediate filament-associated protein in baby hamster (BHK-21) cells. J Cell Biol 1985; 100:620-631.
43. Wiche G, Krepler R, Artlieb U et al. Identification of plectin in different human cell types and immunolocalization at epithelial basal cell surface membranes. Exp Cell Res 1984; 155:43-49.
44. Hieda Y, Nishizawa Y, Uematsu J et al. Identification of a new hemidesmosomal protein, HD1: a major, high molecular mass component of isolated hemidesmosomes. J Cell Biol 1992; 116:1497-1506.
45. Hynes RO. Integrins: a family of cell surface receptors. Cell 1987; 48:549-554.
46. Ruoslahti E, Pierschbacher MD. New perspectives in cell adhesion: RDG and integrins. Science 1987; 238:491-497.
47. Springer TA. Adhesion receptors of the immune system. Nature 1990; 346:425-434.
48. Ruoslahti E. Integrins. J Clin Invest 1991; 87:1-5.
49. Hemler ME, Crouse C, Sonnenberg A. Association of the VLA α6 subunit with a novel protein. J Biol Chem 1989; 264:6529-6535.
50. Kajiji S, Tamura RN, Quaranta V. A novel integrin (αeβ4) from human epithelial cells suggests a fourth family of integrin adhesion receptors. EMBO J 1989; 8:673-680.
51. De Luca M, Tamura RN, Kajiji S et al. Polarized integrin mediates human keratinocyte adhesion to basal lamina. Proc Natl Acad Sci USA 1990; 87:6888-6892.
52. Suzuki S, Naitoh Y. Amino acid sequence of a novel integrin β4 subunit and primary expression of the mRNA in epithelial cells. EMBO J 1990; 9:757-763.
53. Sonnenberg A, Linders CTJ, Daams JH et al. The α6β1 (VLA-6) and α6β4 protein complexes: tissue distribution and biochemical properties. J Cell Sci 1990; 96:207-217.
54. Carter WG, Kaur P, Gil SG et al. Distinct functions for integrins α3β1 in focal adhesions and α6β4/bullous pemphigoid antigen in a new stable anchoring contact (SAC) of keratinocytes: relation to hemidesmosomes. J Cell Biol 1990; 111:3141-3154.
55. Stepp MA, Spurr-Michaud S, Tisdale A et al. α6β4 integrin hetrodimer is a component of hemidesmosomes. Proc Natl Acad Sci USA 1990; 87:8970-8974.
56. Jones JCR, Kurpakus MA, Cooper HM et al. A function for the integrin α6β4 in the hemidesmosome. Cell Regul 1991; 2:427-438.
57. Sonnenberg A, Calafat J, Janssen H et al. Integrin α6/β4 complex is located in hemidesmosomes, suggesting a major role in epidermal cell-basement membrane adhesion. J Cell Biol 1991; 113:907-917.
58. Lee EC, Lotz MM, Steele GD Jr. et al. The integrin α6β4 is a laminin receptor. J Cell Biol 1992; 117:671-678.
59. Burgeson RE, Chiquet M, Deutzmann R et al. A new nomenclature for lamins. Matrix Biology 1994; 14:209-211.
60. Verrando P, Hsi BL, Yeh CJ et al. Monoclonal antibody GB-3, a new probe for the study of human basement membranes and hemidesmosomes. Exp Cell Res 1987; 170:116-128.
61. Rousselle P, Lunstrum GP, Keene DR et al. Kalinin: an epithelium-specific basement membrane adhesion molecule that is a component of anchoring filaments. J Cell Biol 1991; 114:567-576.
62. Carter WG, Ryan MC, Gahr PJ. Epiligrin, a new cell adhesion ligand for integrin α3β1 in epithelial basement membranes. Cell 1991; 65:599-610.
63. Langhofer M, Hopkinson SB, Jones JCR. The matrix secreted by 80G cells contains laminin-related components that participate in hemidesmosome assembly in vitro. J Cell Sci 1993; 105:753-764.
64. Marinkovich MP, Lunstrum GP, Burgeson RE. The anchoring filament protein kalinin is synthesized and secreted as a high molecular weight precursor. J Biol Chem 1992; 267:17900-1796.
65. Kallunki P, Sainio K, Eddy R et al. A truncated laminin chain homologous to the B2

chain: structure, spacial expression, and chromosomal assignment. J Cell Biol 1992; 119:679-693.

66. Vailly J, Verrando P, Champliad, M-F et al. The 100-kDa chain of nicein/kalinin is a laminin B2 chain variant. Eur J Biochem 1994; 219:209-218.
67. Gerecke DR, Wagman DW, Champliaud, M-F et al. The complete primary structure for a novel laminin chain, the laminin B1k chain. J Biol Chem 1994; 269:11073-11080.
68. Baudoin C, Galliano MF, Verrando P et al. The 150 kD subunit of the basement membrane component BM600/nicein shares significant homologies with the chain A of human laminin. J Invest Dermatol 1994; in press.
69. Domloge-Hultsch N, Gammon WR, Briggaman RA et al. Epiligrin, the major human keratinocyte integrin ligand, is a target in both an acquired autoimmune and an inherited subepidermal blistering disease. J Clin Invest 1992; 90:1628-1633.
70. Marchisio PC, Cremona O, Savioia P et al. The basement membrane protein BM-600/nicein codistributes with kalinin and the integrin $\alpha6\beta4$ in human cultured karatinocytes. Exp Cell Res 1993; 205:205-212.
71. Niessen CM, Hogervorst F, Jaspars LH et al. The $\alpha6\beta4$ integrin is a receptor for both laminin and kalinin. Exp Cell Res 1994; 211:360-367.
72. Sonnenberg A, de Melker AA, Martinez de Velasco AM et al. Formation of hemidesmosomes in cells of a transformed murine mammary tumor cell line and mechanisms involved in adherence of these cells to laminin and kalinin. J Cell Sci 1993; 106:1083-1102.
73. Weitzman JB, Pasqualini R, Takuda Y et al. The function and distinctive regulation of the integrin VLA-3 in cell adhesion, spreading, and homotypic cell aggregation. J Biol Chem 1993; 268:8651-8657.
74. Rousselle P, Aumailley M. Kalinin is more efficient than laminin in promoting adhesion of primary keratinocytes and some other epithelial cells and has a different requirement for integrin receptors. J Cell Biol 1994; 125:205-214.
75. Fine, J-D; Smith LT, Holbrook KA et al. The appearance of four basement membrane zone antigens in developing fetal skin. J Invest Dermatol 1984; 83:66-69.
76. Gipson IK, Spurr-Michaud SJ, Tisdale AS. Hemidesmosomes and anchoring fibril collagen appear synchronously during development and wound healing. Devl Biol 1988; 126:253-262.
77. Gipson IK, Grill SM, Spurr SJ et al. Hemidesmosome formation in vitro. J Cell Biol 1983; 97:849-857.
78. Krawczyk WS. A pattern of epidermal cell migration during wound healing. J Cell Biol 1971; 49:247-263.
79. Krawczyk WS, Wilgram GF. Hemidesmosome and desmosome morphogenesis during epidermal wound healing. J Ultrastruct Research 1973; 45:93-101.
80. Briggaman RA, Dalldorf FG, Wheeler CE Jr. Formation and origin of basal lamina and anchoring fibrils in adult human skin. J Cell Biol 1971; 51:384-395.
81. Takahashi Y, Mutasim DF, Patel HP et al. The use of human pemphigoid autoantibodies to study the fate of epidermal basal cell hemidesmosomes after trypsin dissociation. J Invest Dermatol 1985; 85:309-313.
82. Poumay Y, Leclercq-Smekins M, Grailly S et al. Specific internalization of basal membrane domains containing the integrin $\alpha6\beta4$ in dispase-detached cultured human keratinocytes. Eur J Cell Biol 1993; 60:12-20.
83. Poumay Y, Roland IH, Leclercq-Smekins M et al. Basal detachment of the epidermis using dispase: tissue spatial organization and fate of integrin $\alpha6\beta4$ and hemidesmosomes. J Invest Dermatol 1994; 102:111-117.
84. Shellswell GB, Mattey DL, Garrod DR. Hemidesmosome formation by embryonic chick corneal epithelium in vitro. Tiss Cell 1986; 18:659-666.
85. Compton CC, Gill JM, Bradford DA et al. Skin regenerated from cultured epithelial autographs on full-thickness burn wounds from 6 days to 5 years after grafting. Lab Invest 1989; 60:600-612.
86. Croft CB, Tarin D. Ultrastructural studies of wound healing in mouse skin. J Anat 1970; 106:63-77.
87. Hintner H, Fritsch PO, Foidart, J-M et al. Expression of basement membrane zone

antigens at the dermo-epibolic junction in organ cultures of human skin. J Invest Dermatol 1980; 74:200-204.
88. Stock EL, Kurpakus MA, Sambol B et al. Adhesion complex formation after small keratectomy wounds in the cornea. Invest Ophthalmol Vis Sci 1992; 33:304-313.
89. Woodley D, Didierjean L, Regnier M et al. Bullous pemphigoid antigen synthesized in vitro by human epidermal cells. J Invest Dermatol 1980; 75:148-151.
90. Fujikawa LS, Foster CS, Gipson IK et al. Basement membrane components in healing rabbit corneal epithelial wounds: immunofluorescence and ultrastructural studies. J Cell Biol 1984; 98:128-138.
91. Kurpakus MA, Quaranta V, Jones JCR. Surface relocation of alpha6beta4 integrins and assembly of hemidesmosomes in an in vitro model of wound healing. J Cell Biol 1991; 115:1737-1750.
92. Rigal C, Pieraggi, M-T; Vincent C et al. Healing of full-thickness cutaneous wounds in the pig. I. Immunohistochemical study of epidermo-dermal junction regeneration. J Invest Dermatol 1991; 96:77-785.
93. Contard P, Bartel RL, Jacobs II L et al. Culturing keratinocytes and fibroblasts in a three-dimensional mesh results in epidermal differentiation and formation of a basal lamina-anchoring zone. J Invest Dermatol 1993; 100:35-39.
94. Contard P, Jacobs II L, Perlish JS et al. Collagen fibrillogenesis in a three-dimensional fibroblst cell culture system. Cell Tissue Res 1993; 273:571-575.
95. Regauer S, Seiler GR, Barrandon Y et al. Epithelial origin of cutaneous anchoring fibrils. J Cell Biol 1990; 111:2109-2115.
96. Riddelle KS, Green KJ, Jones JCR. Formation of hemidesmosomes in vitro by a transformed rat bladder cell line. J Cell Biol 1991; 112:159-168.
97. Riddelle KS, Hopkinson SB, Jones JCR. Hemidesmosomes in the epithelial cell line 804G: their fate during wound closure, mitosis and drug induced reorganization of the cytoskeleton. J Cell Sci 1992; 103:475-490.
98. Baker J, Garrod D. Epithelial cells retain junctions during mitosis. J Cell Sci 1993; 104:415-425.
99. Spinardi L, Ren, Y-L; Sanders R et al. The β4 subunit cytoplasmic domain mediates the interaction of α6β4 integrin with the cytoskeleton of hemidesmosmes. Mol Biol Cell 1993; 4:871-884.
100. Woodley DT, Burgeson RE, Lunstrum G et al. Epidermolysis bullosa acquisita antigen is the globular carboxyl terminus type VII procollagen. J Clin Invest 1988; 81:683-687.
101. Tatnall FM, Whitehead PC, Black MM et al. Identification of the epidermolysis bullosa acquisita antigen by LH 7,2 monoclonal antibody: use in diagnosis. Br J Dermatol 1989; 120:533-539.
102. Shimizu H, McDonald JN, Gunner DB et al. Epidermolysis bullosa acquisita antigen and the carboxy terminus of type VII collagen have a common immunolocalization to anchoring fibrils and lamina densa of basement membrane. Br J Dermatol 1990; 122:577-585.
103. Woodley DT, Briggaman RA, Gammon WR. Acquired epidermolysis bullosa, a bullous disease associated with autoimmunity to type VII (anchoring fibril) collagen. Dermatol Clin 1990; 8:717-726.
104. Gammon WR. Epidermolysis bullosa acquisita: a disease of autoimmunity to type VII collagen. J Autoimmun 1991; 4:59-71.
105. Epstein EH Jr. Molecular genetics of epidermolysis bullosa. Science 1992; 256:799-804.
106. Bonifas JM, Rothman AL, Epstein EH Jr. Epidermolysis bullosa simplex: evidence in two families for keratin gene abnormalities. Science 1991; 254:1202-1205.
107. Fuchs E, Coulombe PA. Of mice and men: genetic skin diseases of keratin. Cell 1992; 69:899-902.
108. Pulkkinen L, Christiano AM, Airenne T et al. Mutations in the γ2 chain gene (LAMC2) of kalinin/laminin 5 in the junctional forms of epidermolysis bullosa. Nature Genet 1994; 6:293-298.
109. Burgeson RE, Lunstrum GP, Rokosova B et al. The structure and function of type VII collagen. Ann NY Acad Sci 1990; 32-43.
110. Ryynanen M, Knowlton RG, Parente MG et al. Human type VII collagen: genetic

linkage of the gene (COL7A1) on chromasome 3 to dominant dystrophic epidermolysis bullosa. Am J Hum Genet 1991; 49:797-803.

111. al Imara L, Richards AJ, Eady RA et al. Linkage of autosomal dominant dystrophic epidermolysis bullosa in three British families to the marker D3S2 close to the COL7A1 locus. J Med Genet 1992; 29:381-382.

112. Gruis NA, Bavinck JN, Steijlen PM et al. Genetic linkage between the collagen VII (COL7A1) gene and the autosomal dominant form of dystrophic epidermolysis bullosa in two Dutch kindreds. J Invest Dermatol 1992; 99:528-530.

113. Ryynanen M, Ryynanen J, Sollberg S et al. Genetic linkage of type VII collagen (COL7A1) to dominant dystrophic epidermolysis bullosa in families with abnormal anchoring fibrils. J Clin Invest 1992; 89:974-980.

114. Hovnanian A, Duquesnoy P, Blanchet Bardon C et al. Genetic linkage of recessive dystrophic epidermolysis bullosa to the type VII collagen gene. J Clin Invest 1992; 90:1032-1036.

115. Sollberg S, Ryynanen J, Olsen DR et al. Transforming growth factor-β up-regulates the expression of the genes for β4 integrin and bullous pemphigoid antigens (BPAG1 and BPAG2) in normal and transformed human keratinocytes. J Invest Dermatol 1992; 99:409-414.

116. Tamai K, Li K, Uitto J. Identification of a DNA-binding protein (keratinocyte transcriptional protein-1) recognizing a keratinocyte-specific regulatory element in the 230-kDa bullous pemphigoid antigen gene. J Biol Chem 1994; 269:493-502.

117. Kimmel KA, Carey TE. Altered expression in squamous carcinoma cells of an orientation restricted epithelial antigen detected by monoclonal antibody A9. Cancer Res 1986; 46:3614-3623.

118. Carey TE, Wolf GT, Hsu S et al. Expression of A9 antigen and loss of blood group antigens as determinants of survival in patients with head and neck squamous carcinoma. Otolaryngol Head Neck Surg 1987; 96:221-230.

119. Wolf GT, Carey TE, Schmaltz SP et al. Altered antigen expression predicts outcome in squamous cell carcinoma of the head and neck. J Natl Cancer Inst 1990; 82:1566-1572.

120. Van Waes C, Kozarsky KF, Warren AB et al. The A9 antigen associated with aggresive human squamous carcinoma is structurally and functionally similar to the newly defined integrin α6β4. Cancer Res 1991; 51:2395-2402.

121. Carey TE, Laurikainen L, Nair TS et al. Regulation of expression and phosphorylation of A9/α6β4 integrin in normal and neoplastic keratinocytes. J Natl Cancer Inst Monogr 1992; 13:75-86.

122. Carey TE, Nair TS, Chern C et al. Blood group antigens and integrins as biomarkers in head and neck cancer: is aberrant tyrosine phosphorylation the cause of altered α6β4 integrin expression? J Cell Biochem 1993; 17F:223-232.

123. Savoia P, Trusolino L, Pepino E et al. Expression and topography of integrins and basement membrane proteins in epidermal carcinomas: basal but not squamous cell carcinomas display loss of α6β4 and BM-600/nicein. J Invest Dermatol 1993; 101:352-358.

124. Sager R, Anisowicz A, Neveu M et al. Identification by differential display of alpha 6 integrin as a candidate tumor supressor gene. FASEB J 1993; 7:964-970.

125. Hynes RO. Integrins: versatility, modulation, and signaling in cell adhesion. Cell 1992; 69:11-25.

126. Gospodarowicz D. In: Barnes DW, Sirbasku DA, Stao GH, eds. Methods for Preparation of Media, Supplements, and Substrata for Serum-Free Animal Cell Culture. Vol. 1. New York: Alan R Liss Inc, 1984:275-293.

127. Hashimoto T, Masayuki A, Ebhara T et al. Further analysis of epitopes for human monoclonal anti-basement membrane zone antibodies produced by stable human hybridoma cell lines constructed with Epstein-Barr virus transformants. J Invest Dermatol 1993; 100:310-315.

CHAPTER 6

Basement Membrane-Hemidesmosome Interactions

John D. Aplin, Heather J. Church

THE OCCURRENCE OF HEMIDESMOSOMES

Hemidesmosomes (HDs) are small (0.1-0.5 μm diameter) electron dense plaques that anchor epithelial cells to the underlying basement membrane.[1-7] HDs are associated with tissues in which resistance to mechanical stress is required for the maintenance of epithelial integrity. Examples include skin, bladder, trachea, cornea, esophagus, amnion, oral mucosa, cervix, vagina, ameloblasts of developing tooth and myoepithelium of breast. Since HDs are lost during wound healing[8,9] they may be assumed to be characteristic of non-motile cells. Other types of epithelial cell can be observed to form related specializations at the basal surface both in vivo and in vitro. These may be smaller than normal HDs or may lack some of their morphological features and it is not yet clear to what extent they resemble HDs at the molecular level. There is another group of epithelial tissues in which no ultrastructural evidence has been found for the presence of plaque-forming junctions at the epithelial basal surface. This group includes trophoblast of the villous placenta, intestinal epithelium and the podocytes of the kidney glomerulus. Non-epithelial cell types that are anchored to basement membrane include muscle, nerve and vascular endothelial cells, and while these may possess submembranous densities, they all lack true HDs as monitored by ultrastructural examination.

MORPHOLOGY OF THE ECM AT THE HEMIDESMOSOME

The role of HDs is to link the intermediate filament cytoskeleton through the basal epithelial plasma membrane to the subjacent basement

Molecular Biology of Desmosomes and Hemidesmosomes, edited by Jane E. Collins and David R. Garrod.

membrane. HDs are recognizable by a set of ultrastructural criteria[1-5] (Fig. 6.1): prominent intermediate filament bundles in the adjacent cytoplasm, a well-formed dense plaque at the inner face of the plasma membrane, a sub-basal dense plate, anchoring filaments running through the lamina lucida, a lamina densa and underlying anchoring fibrils. Not all of these features may be readily apparent in all planes of section; the sub-basal dense plate is only observed in near-perpendicular sections,[10] visualization of anchoring filaments is dependent on the method of specimen preparation (see below) and anchoring fibrils are not seen in all planes of section, perhaps because of their curvature.

Very prominent, uniformly-oriented anchoring fibrils can be observed in axolotl skin[4] in which it is possible to follow a colinear array of fibrillar structures upwards from the anchoring fibril through the lamina densa and along anchoring filaments to the basal cell membrane. On this basis it has been proposed that there may be a continuous structure linking the stromal ECM to intermediate filaments in the epithelial cytoplasm in other tissues and species, but this not often as clearly demonstrable as in amphibian skin.

Figure 6.1 shows the basal epithelial surface and basement membrane of human amnion. All of the above features are present. The HD plaques are well formed and are often most prominent at or near the distal ends of basal cell foot processes. Anchoring filaments traverse the lamina lucida at the HD, and the lamina densa is more electron dense at these sites. Anchoring fibrils are not as numerous in this tissue as in skin, but may readily be observed. One of the most important functions of amnion epithelial cells is deposition of stromal interstitial extracellular matrix beneath the basal lamina.[5,11,12] This is also the case in corneal epithelium.[13] There are two quite discrete areas of basal plasma membrane in amnion cells: the groups of HDs on the foot processes and the basal cell invaginations opposite which the lamina densa is significantly looser, and within which can often be found stromal-type fibrillar or microfibrillar ECM. All these components can also be observed in the underlying stroma. It has been suggested that the basal cell invaginations are repositories of newly secreted ECM containing both stromal and basal laminal components and that organization into spatially discrete supramolecular associations follows secretion into the invaginations.[5,11,12]

Essentially all basement membranes examined using conventional electron microscopic preparation methods contain readily distinguishable laminas lucida and densa (Fig. 6.1). However, recent observations made using novel preparation methodology suggest that this distinction is an artifact of dehydration. Basal lamina observed after fast freezing and freeze-substitution consists of a fairly uniform, finely fibrillar or amorphous continuum from the basal plasma membrane down to the basal edge of the lamina.[14,15] In these preparations the anchoring filaments are not visible.

BASEMENT MEMBRANE COMPOSITION

Basement membranes contain a scaffolding of collagen IV (most frequently $\alpha1(IV)_2\alpha2(IV)$) that assembles into a continuous and flexible network of fine fibrils or cords (1.5-4 nm diameter). The flexibility of the assembly arises as a result of interruptions in the GXY triplet repeat sequence that is characteristic of collagenous proteins: overlapping interruptions in the subunits produce about 26 interruptions of variable length in the triple helical monomer. Network assembly occurs by self-association of monomer into dimer at the C-terminal end and tetramer at the N-terminal end, and further side-to-side interactions also occur to generate a three dimensional meshwork in which the intermolecular spaces can vary in size.[16] This network is stabilized by disulphide bonding as well as by hydroxylysine-derived cross-links.

Collagen IV has emerged as a family of isoforms in which different gene products can combine to form heterotrimeric molecules with different subunit permutations.

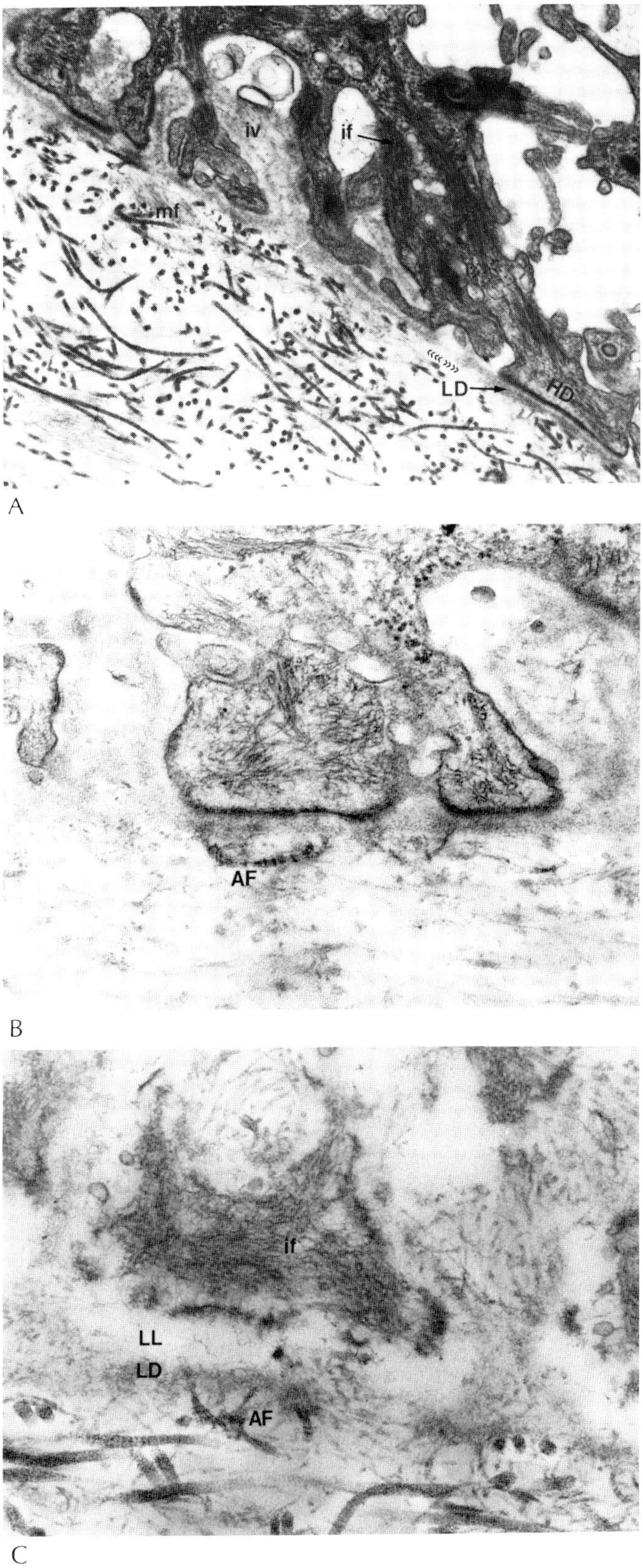

Fig. 6.1. Transmission electron micrographs of the basement membrane zone of human term amnion epithelium. (A) The basal cell surface is highly invaginated. The cell processes that extend downwards towards the basal lamina are rich in intermediate filament bundles (if). HDs are associated with the more distal parts of the processes. The lamina densa (LD) is rather more electron dense at the sites of HDs; elsewhere it has evident discontinuities (<<>>), and within the invaginations (iv) the lamina densa becomes more loosely organized. Microfibril bundles (mf) are closely associated with the stromal side of the lamina densa. Striated collagen fibrils are abundant in the stroma. x38,000. (B) Higher power view (x75,000) showing an anchoring fibril (AF) looping through the ECM beneath the lamina densa. This probably represents a parallel aggregate of several of the dimeric species shown in Fig. 6.3B. Note that the intermediate filaments (if) do not make direct contact with the dense plaque, but instead pass through the adjacent cytoplasm. (C) A detergent-extracted preparation[39] in which plasma membrane has been solubilized from areas not associated with HDs. An area is shown that is similar to the unextracted preparation in b. The lamina lucida (LL) contains anchoring filaments which pass from the lamina densa (LD) to the plasma membrane at the HD dense plaque. Note that the ECM beneath the lamina densa contains large striated collagen fibrils as well as microfibrils. The latter run directly into the lamina densa.

Present data suggest that the most widely occurring isoform $\alpha1(IV)_2\alpha2(IV)$ is present in tissues where HD formation occurs. The occurrence of other subunits is incompletely documented; α3(IV), α4(IV) and α5(IV) are found in kidney, and α3(IV) and α4(IV) are in the synaptic basement membranes of muscle and in Bowman's capsule of the eye.[17] There is no current evidence to link these isoforms to HDs.

Another basement membrane component that is known to be capable of self-polymerization is the large glycoprotein laminin. Laminin is a ubiquitous component of basement membranes and its structure is presented in detail in a later section. Current models of the molecular disposition of laminin in basement membranes are based on the isoform laminin 1 isolated from the mouse EHS tumor. It has been suggested[16] that laminin 1 polymer could interweave within the network of collagen IV cords, and the assembly could be stabilized by entactin (also known as nidogen), a glycoprotein of 150 kDa that exhibits binding sites for both collagen IV and laminin 1.[18,19] Another possibility is that laminin 1 monomer simply binds directly to the collagen IV network.[20,21]

Like collagen IV, laminin exists in different isoforms made up of three subunits permuted from a family.[22] Isoforms of laminin may play very specific roles in the attachment of cells via HDs and these isoforms show corresponding tissue-specific expression. This is discussed below.

One other component of basement membranes appears to be ubiquitous. The heparan sulphate proteoglycan known as perlecan[23] binds to laminin, collagen IV, entactin and itself.[21,24] The interaction with collagen IV has in principle the potential to limit lateral association of fibrils. Perlecan binding can also promote or destabilize laminin polymerization at different concentrations.[16] Alternatively spliced forms of the perlecan core polypeptide have been detected.[23,25]

Other components of the basement membrane occur with varying abundance[23,26] in basement membranes. These include BM40/SPARC/osteonectin, BM90, amyloid P, thrombospondin and components with tissue-specific functions such as acetylcholine esterase and agrin. None of these species is consistently present in the major sites of HD formation and it appears unlikely that any of them plays a fundamental role in HD assembly or architecture.

Beneath the lamina densa, anchoring fibrils containing collagen type VII are found (Fig. 6.1; see below). In addition there are often accumulations of microfibrillar components that represent in skin the 'terminal arbourisations of the dermal elastic system[4,10] (Fig. 6.1). Independent microfibrillar systems are formed by collagen type VI[27] and fibrillin.[28] The chondroitin sulphate proteoglycan versican is associated with microfibrils and it is enriched near the dermal-epidermal junction.[29]

Thus while it is clear that all basement membranes share structural components, including type IV collagen, laminin, and heparan sulphate proteoglycan, there is an intriguing compositional heterogeneity.[30,31] This can arise in a number of ways: some basement membranes contain unique polypeptides not found in other tissues; tissue-specific glycosylation of basement membrane components may give rise to novel glycoforms and epitopes; and both collagen IV and laminin exist as a family of isoforms arising by combining the products of several unique genes in sets of three. In addition, the superficial layers of stroma are likely to be distinct in different tissues.

Some of these variations may be explained by the fact that the origin of basement membrane also varies. Thus, for example, in skin, dermal fibroblasts contribute macromolecules to the basement membrane,[32] while in late gestation amnion there are no fibroblasts found for several microns beneath the epithelium and it is the epithelial cells that produce both the basement membrane and superficial stroma.[5,12]

MOLECULAR REQUIREMENTS FOR HEMIDESMOSOME FORMATION

The molecular composition of the HD has not yet been completely defined. However, constituents that localize to specific structures within the HD have been recognized: the dense plaque polypeptides BPAG230 and HD1,[33-36] the plasma membrane glycoproteins integrin $\alpha6\beta4$[37-39] and BPAG180,[40-42] laminin 5 in anchoring filaments,[43] and collagen VII in anchoring fibrils.[44] There are also less well-characterized components such as the 19-DEJ-1 epitope, which localizes to the lamina lucida of HDs[45] and a 125 kDa anchoring fibril-associated component.[46] These are some of the building blocks required to produce the mature HDs of skin, amnion or cornea. However, the same proteins are not excluded from non-HD assemblies. Thus, for example, very prominent anchoring fibrils may be observed in rat vas deferens despite the absence of HDs.[47] Similarly, integrin $\alpha6\beta4$ is a surface component of many different cells including vascular endothelium[48-50] and trophoblast of villous placenta,[51] neither of which form HDs. Conversely, a few HD-like structures (though not true HDs as defined above) have been observed that lack one or more of these molecular components. Thus the epithelial meninges of brain and Purkinje fibers in the heart exhibit HD-like specializations but lack BPAG230.[33] However, these cells appear to be exceptions to the general rule.

A true test of the function of HD constituents occurs in pathological situations where absence or abnormality of a specific component leads to the breakdown of the tissue interface at the site of the HD. This occurs in the blistering disease epidermolysis bullosa[52] and various forms of the disease have already given confirmation of the importance of collagen VII and laminin 5 (see below) as well as intracellular components of the HD in epithelial cell adhesion at specific tissue sites.

THE LAMININ FAMILY

The prototypical laminin (laminin 1) was originally described in the matrix of the Engelbreth-Holm-Swarm (EHS) mouse tumor,[53] and on isolation was found to be a large heterotrimeric glycoprotein (> 800 kDa). The three subunits $\alpha1$ (A), $\beta1$ (B1) and $\gamma1$ (B2) assemble to yield a cruciform molecule with one long arm and three short arms (Fig. 6.2). The long arm of the cross comprises a linear assembly of α-helical domains (about 600 residues) associated into a triple helix. At the end of the long arm is a series of globular domains G1-5 arising from the C-terminal region of the $\alpha1$ chain. Each short arm arises from the N-terminal region of a single subunit polypeptide. The molecule has 74 potential N-glycosylation sites unevenly distributed between the chains. Many of these sites are concentrated in the long arm.[54-56]

It has now become clear that laminin 1 is a member of a family of related molecules, each a disulphide linked heterotrimer containing an α, β and γ chain. The homologous chains are products of different genes.[55] Isoforms are shown in Table 6.1 with both original and current nomenclatures.[22] Table 6.2 shows the various subunits with their molecular weights and sequence identity to the $\alpha1$, $\beta1$ and $\gamma1$ chains.[55,57] Note that not all the possible subunit combinations have as yet been observed, and it is possible that new laminin isoforms remain to be discovered.

Early immunolocalization studies used polyclonal antisera that recognized $\alpha1$, $\beta1$ and $\gamma1$ chains as a result of which it was thought that laminin 1 was ubiquitous in basement membranes. However, it has become apparent that different laminin isoforms are expressed in different tissues and at different stages of development. Laminin 2 (merosin) was originally isolated from human placenta[58] and mouse heart.[59,60] It was found to contain both a $\beta1$ and a $\gamma1$ chain, along with an α chain homologue of 300 kDa. An 80 kDa polypeptide was also associated with this α chain. Laminin 3 (s-laminin) was found to contain a homologous $\beta2$ chain and to occur in synaptic and glomerular basement membrane.[61]

The development of subunit specific monoclonal antibodies has been important in generating information as to the tissue distribution of these new laminin subunits. It has been found that their expression is limited and often (but not always) exclusive. An example is the occurrence of both laminins 1 and 2 in heart. Laminin 1 is restricted to the basement membrane of blood vessels, while laminin 2 is expressed only in the cardiac myocyte basement membrane.[55] In general, laminin 1 is widespread in epithelia and endothelia. Laminin 2 is found in striated muscle, peripheral nerve and trophoblast and laminin 3 occurs at nerve synapse and kidney glomerulus. Laminin 4 (s-merosin) has been identified in myotendinous junction and trophoblast.[62,63] Laminin 5 has been detected in the epithelium of skin, amnion, trachea, esophagus and bladder, but it is absent from liver and kidney.[64] Laminins 6 and 7 are the most recently discovered members of the family.[65]

Laminin 1 is readily available from the EHS tumor from which it can be extracted with neutral buffers and EDTA, typically yielding 100 mg from 50 g starting material.[53,66] In contrast, the extraction of native laminins from normal tissue has proved to be more difficult. As a consequence most of the structural and functional studies have been carried out on laminin 1.

In vitro studies have shown that laminin 1 is able to promote cell growth, adhesion and differentiation. Adhesion has been shown to be mediated by members of the integrin family[67-69] (see below). There is also the possibility of carbohydrate moieties on laminin acting as recognition sites for lectin-like receptors.[67]

The differences in cell binding activities of laminin isoforms are less clear. Comparison of rat L2 yolk sac tumor laminin 1 and human placental laminin 2 in terms of cell attachment, spreading and neurite promotion in a range of cells failed to show any differences.[55] Adhesion appeared to be mediated by the β1 integrins. There is some indication that laminin 3 may have unique properties based on the observation that regenerating nerves form contacts with muscle at previously existing synapse sites. The β2 subunit has been found to contain unique sequences that can influence the attachment of neurons, and it may be that the neuromuscular junction can only regenerate at a site containing laminin 3.[70] There is clear evidence that laminin 5 differs from laminin 1 in its ability to support cell adhesion and this is elaborated in a later section.

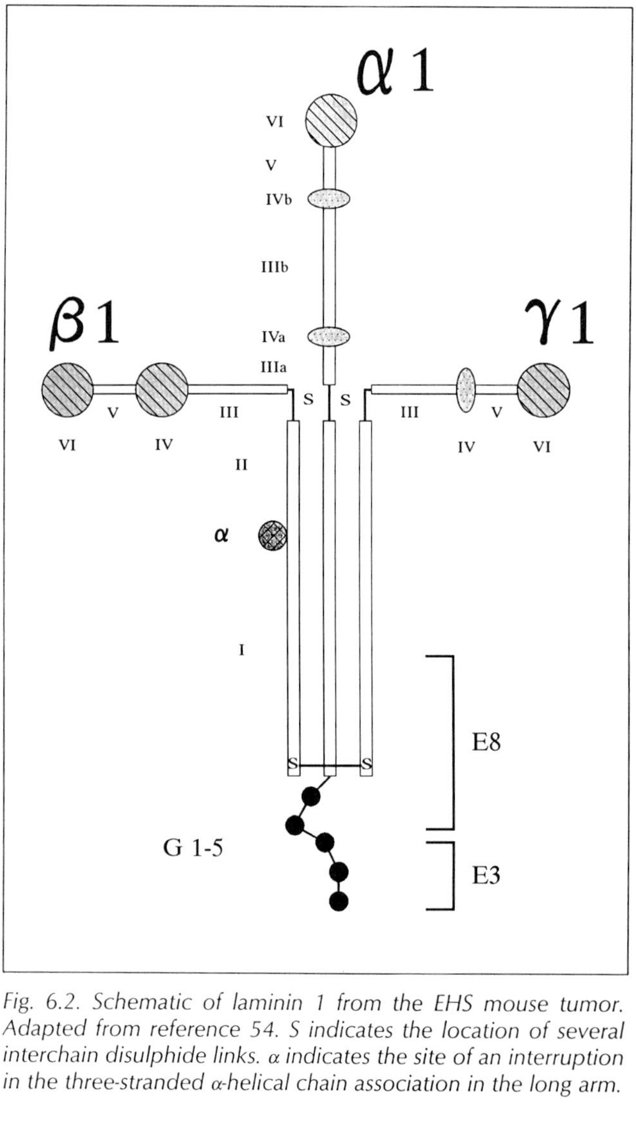

Fig. 6.2. Schematic of laminin 1 from the EHS mouse tumor. Adapted from reference 54. S indicates the location of several interchain disulphide links. α indicates the site of an interruption in the three-stranded α-helical chain association in the long arm.

LOCALIZATION OF LAMININ ISOFORMS AT HEMIDESMOSOMES

Laminin 5 (kalinin/nicein) was originally identified independently by different laboratories. Kalinin was localized to anchoring filaments in skin HDs by immunogold labeling and electron microscopy.[43] It is secreted by keratinocytes in culture as a precursor protein consisting of three disulphide-bonded polypeptides (200, 155 and 105 kDa) which undergo extracellular processing to yield subunits of 165, 140 and 105 kDa respectively.[71] These make up the mature tissue form. Examination by rotary shadowing electron microscopy showed kalinin to be a rod-like molecule with globular ends.[65] Immunolocalization studies of kalinin indicated that it is restricted to the basement membrane zone of skin, amnion, cornea, trachea and esophagus.[43]

Table 6.1. Laminin heterotrimers. Note that not all possible chain combinations have yet been observed

Name	Previous Name	Chain Composition
LAMININ 1	EHS LAMININ	$\alpha1\beta1\gamma1$
LAMININ 2	MEROSIN	$\alpha2\beta1\gamma1$
LAMININ 3	S-LAMININ	$\alpha1\beta2\gamma1$
LAMININ 4	S-MEROSIN	$\alpha2\beta2\gamma1$
LAMININ 5	KALININ/NICEIN	$\alpha3\beta3\gamma2$
LAMININ 6	K-LAMININ	$\alpha3\beta1\gamma1$
LAMININ 7	K-S-LAMININ	$\alpha3\beta2\gamma1$

Nicein is also a disulphide-linked heterotrimer with subunit molecular weights comparable to those of kalinin. It localizes to the anchoring filaments in skin HDs.[64,72] Biochemical analysis has indicated that the three kalinin and nicein subunits are identical.[73] Cloning of the corresponding human cDNAs demonstrated sequence homology with laminin 1 in all 3 subunits,[74] and the molecule has been renamed laminin 5 with subunit designations α3 (165 kDa), β3 (140 kDa) and γ2 (105 kDa). The corresponding genes (Table 6.2) map to chromosomes 18 (LAMA3) and 1 (LAMB3 and LAMC2) in both human[75] and mouse.[76] Human and mouse laminin 5 are highly conserved, showing 85-90% identity.[77] The α3 chain shows only 20% identity with the α1 chain,[78] while the β3 chain is 40% identical to the β1[79] and the γ2 and γ1 chains are 45% identical.[80] One specific feature of the γ2 chain is the absence of sequences corresponding to domains IV and V in the γ1 chain (Fig. 6.2). Domain VI however is retained in γ2. Laminin B2t, which was initially identified as a truncated form of the γ1 chain,[81] has now been identified as γ2.[80] To date only 2 γ chains have been described.

Human laminin 5 is expressed in specialized epithelia from the 7th week of gestation,[82] which is consistent with the time of initial appearance of HDs and type VII collagen in embryonic skin.[83] In situ hybridization and immunological techniques have been used to establish the expression pattern of all three subunits of

Table 6.2. Subunits of the laminin family with former and current nomenclature

New Name	Previous Name	Sequence Identity*	Molecular Weight	Gene
α1	A, Ae	-	440 kDa	LAMA1
α2	M, Am	40%	340	LAMA2
α3	Kal-165, Ak	20%	200/165	LAMA3
β1	B1, B1e	-	220	LAMB1
β2	S, B1s	50%	180	LAMB2
β3	Kal-140, B1k	40%	140	LAMB3
γ1	B2, B2e	-	200	LAMC1
γ2	Kal-105 B2k, B2t	45%	155/105	LAMC2

* - Identity with the corresponding laminin 1 subunits.

laminin 5 during mouse development.[77] Significant levels of expression were found during organogenesis of respiratory and digestive systems and in the enamel-secreting ameloblasts of developing tooth. Consistent with other sites of expression, these latter cells exhibit HDs at their interface with enamel.[84] The β3 and γ2 subunits, but not the α3 subunit, were observed in chorioid plexus. This suggests the existence of a heterodimer lacking α3, or a heterotrimer containing a different α chain. Thus the β3 and γ2 chains may play a specific role in developing CNS. The α3 subunit is expressed in the floor plate of the neural tube and may be involved in the migration of motor neurons.[77]

Purification of laminin 5 from amnion resulted in the isolation of a complex in which the laminin 5 is covalently bonded to laminin 6 (k-laminin), and this complex is also produced by keratinocytes in culture.[85] Laminin 6 contains a β1 and a γ1 chain in association with a novel α chain (195 kDa). Monoclonal antibody BM165 to the α3 chain also recognized this chain, although owing to the discrepancy in size, the extent of their similarity is not clear. Rotary shadowing has demonstrated that the short arm observed in laminin 1 that corresponds to the α1 chain N-terminal domain has no analog in laminin 6.[65] The distribution of laminin 6 is not yet known in detail.

Epiligrin is a protein also produced by cultured keratinocytes that has been demonstrated to support cell adhesion.[86] Monoclonal antibodies P1E1 to epiligrin and GB3 to laminin 5 reveal the same tissue distribution of antigen and radioimmunoprecipitate the same polypeptides.[87] However, since more than three subunits were recovered, the precise relationship between epiligrin and laminin 5 is not clear. Epiligrin may be a complex of laminin 5 with another protein, possibly laminin 6.[80]

LAMININ-5 IS INVOLVED IN CELL ADHESION IN SKIN

The first evidence of involvement of laminin 5 in cell adhesion was obtained by the use of the monoclonal antibody BM165 which releases cultured keratinocytes from their substrate.[43] Antibodies to laminin 1, collagens IV or VII could not cause detachment. Fibroblasts treated with antibody BM165 did not detach from the substrate. Ultrastructural examination showed primitive HD-like densities at the basal surface of the keratinocytes. In the presence of BM165, keratinocytes also failed to adhere to matrix remaining after cell detachment with EDTA, nor could they attach to uncoated plastic. Similarly the incubation of skin fragments with this antibody resulted in detachment of the epidermis from the dermis through the lamina lucida.[43] In this experiment α3 immunoreactivity was retained on the dermal side of the cleavage plane, in the area that would have been the lamina lucida. This is a clear demonstration that the α3 chain, or a closely associated component, is required for epidermal anchorage to the basement membrane. It also suggests that laminin 5 may be involved in this anchorage.

Further indication that laminin 5 is essential for normal dermal-epidermal adhesion is the clinical pathology of the Herlitz form of junctional epidermolysis bullosa (JEB). This is a severe autosomal blistering disease characterized by structurally abnormal and greatly reduced HDs, blister cleavage through the lamina lucida[88] and absent or abnormal expression of laminin 5.[89,90] The distribution of laminin 5 correlates well the disease; affected tissues include skin, oral cavity including teeth,[91] esophagus and larynx. Epiligrin appears to be absent or weak in JEB skin.[87] There are also modifications to the integrin β4 chain ectodomain in JEB skin, with the loss of an epitope recognized by monoclonal antibody G71, but other β4 epitopes are conserved and so this may be a secondary effect resulting, for example, from proteolysis.[50]

JEB has given important insights into the role of laminin 5 in basement membranes. It was first implicated by immunolocalization studies using monoclonal antibody GB3 which binds laminin 5 and was found to be unreactive with the dermal-epidermal junction in JEB.[92] However, since the GB3 epitope has not been defined further, the expression of the individual subunits was not established. Immunolocalization studies using subunit-specific antibodies have suggested that one subunit can be absent while the other two are present.[93,94] For example, some patients lack the β3 subunit while the α3 and γ2 subunits are present both in the basal epidermal cytoplasm and basement membrane zone. Patients lacking the α3 or the γ2 chain have also been identified.[94] Northern blotting studies have demonstrated absence of the corresponding mRNAs, suggesting either a reduction in transcription, or mRNA instability.

These results suggest that JEB arises from impaired production of mature laminin 5 due to mutations in any one of the three subunits, and that the condition is genetically heterogeneous. Mutations in the LAMC2 gene have been identified in JEB families.[95,96] In one family a homozygous point mutation leading to a premature termination codon was identified, while in another an insertion-deletion mutation resulted in a frameshift also giving a premature termination codon. In both of these cases a truncated peptide is predicted. In some families linkage to the LAMC2 gene has been excluded, implying that mutations may occur at the LAMA3 or LAMB3 loci.[57]

CELL INTERACTION WITH LAMININ-5 IN VITRO

The formation of HDs by epithelial cells in vitro is not a general phenomenon, even when the cells have the capability to produce them in vivo.[5] It is likely that there is a requirement for a specific substrate composition or ECM to allow the production of HDs; the necessary components may of course be deposited by the adherent cells. The rat bladder carcinoma cell line 804G forms HD-like structures in vitro which contain the plaque component BPAG230[97] and the HD cell surface component integrin α6β4.[98] HD assembly could be inhibited by plating cells in the presence of antiserum to integrin α6β4.[98] This implicates α6β4 in mediating HD formation. 804G cells produce an ECM that directs the formation of HD-like structures containing BPAG230, BPAG180 and integrin α6β4 by a human epidermal carcinoma cell line SCC12.[99] Characterization of the 804G ECM identified a major component as a heterotrimeric glycoprotein containing subunits of 150, 140 and 135 kDa. The 140 kDa subunit shows 80% homology with the γ2 chain suggesting that the 804G ECM may contain rat laminin 5.[99]

The murine mammary tumor cell line RAC-11P/SD, which differentiates into squamous epithelium and produces HD-like structures in vitro, deposits an ECM that contains laminins 1 and 5, type IV collagen and fibronectin.[100] Laminins 1 and 5 both concentrate beneath HD-like specializations in these cultures and they colocalize with integrin α6β4. When RAC-11P/SD cells were plated on exogenous laminins 1 or 5, attachment occurred in each case but cell spreading occurred on laminin 5 and not laminin 1. Attachment could not be blocked by antibody to the integrin α6 subunit. Attachment was followed by redistribution of α6β4 to the basal cell surface. If integrin redistribution was prevented using metabolic inhibitors, attachment became sensitive to inhibition by antibody to the integrin α6 subunit.

In healing cornea, migrating epithelial cells deposit basement membrane components including laminin and collagen type VII onto the connective tissue surface and HDs form in the healing epithelium. Polyclonal antibodies to integrin β4 both inhibit HD formation and induce cell detachment where HDs have already formed. Monoclonal antibodies to integrin α6 subunit also detach the epithelium.[101]

These data suggest that cell adhesion and migration on a substratum and HD assembly are distinct processes. However, it is important to underline the difficulty of studying the requirements for HD assembly in living cells in which sequestration of cell surface components into basal domains may inhibit antibody access.

Experiments have been conducted in cell lines that do not form HDs to investigate possible binding interactions between integrin α6β4 and laminin isoforms at cell attachment. Interpretation of the data is complicated by the fact that laminins 1 and 5 have been shown to interact with several members of the integrin family, and most cells express several integrins. Thus laminin 1 binds integrins α1β1, α2β1, α3β1, α6β1, α7β1, α6β4 and αvβ3.[102] Cell attachment to laminin 5 may be mediated by α3β1, α6β1[103] or α6β4.[104] However, no integrin β1 immunoreactivity can be detected in skin HDs,[38] so it is assumed that adhesion mediated by this integrin subfamily occurs in other contexts such as wound healing.

The intestinal carcinoma cell line Clone A uses α6β4 to bind laminin 1,[105] but other α6β4-bearing cell lines either show poor adhesion or use alternate receptors.[103,106,107] The use of transfected human erythroleukemic K562 cells expressing both α6β1 and α6β4 has provided a useful experimental model for interactions between the α6 integrins and laminin isoforms. The α6 subunit exists in 2 variants, α6A and α6B, which have different cytoplasmic domains resulting from alternate mRNA splicing. Initially transfected with α6A or α6B, heterodimeric complexes containing α6Aβ1 or α6Bβ1 were displayed on the cell surface. The cells were then (on activation with a monoclonal antibody to the integrin β subunit) able to bind laminin 1 (via its E8 domain) or laminin 5. Adhesion to either laminin 1 or laminin 5 could be blocked by anti-α6 monoclonal antibody (GoH3). Thus α6β1 could mediate adhesion to laminin 1 or laminin 5.[108]

Cells further transfected with an integrin β4 construct expressed α6Aβ4 or α6Bβ4 in addition to the corresponding α6β1 complexes. In this case it was found that adhesion to either laminin 1 or laminin 5 could be blocked by anti-α6 antibody (GoH3). Antibody to the β4 subunit produced a small reduction in adhesion to laminin 5, and a more significant inhibition of adhesion to laminin 1. Interpretation of these data is hindered by the presence of more than one interacting integrin at the cell surface. However, solubilized integrin α6β4 bound to ECM containing laminins 1 or 5 in the presence of Mg^{2+} ions and this interaction could be inhibited by antibody GoH3 to α6.[104]

Epiligrin can act as a cell adhesion receptor for keratinocytes in vitro, but this interaction is blocked by anti-integrin β1 antibody and is thought to be mediated by integrin α3β1.[86] However, epiligrin is also found to colocalize with integrin α6β4 and BP antigens in vitro. On detergent extraction of cell monolayers, α6β4 and BP180 remain complexed with epiligrin suggesting the possibility of a stable HD-type contact.[86] Integrin α6β4 redistributes to colocalize with laminin 5 in keratinocytes in vitro within 6 hours of plating.[109] These studies suggest that while initial adhesion may be mediated by integrin α3β1, more stable adhesion complexes may be mediated by α6β4 interaction with laminin 5 which may in turn support the development of HD-like structures.

In summary, data gathered in vitro indicates that the HD component integrin α6β4 can interact with laminin 5 or laminin 1 although the former interaction is difficult to demonstrate in cell attachment assays. However, the fact that JEB skin is characterized by very poorly developed HDs despite normal expression of laminin 1, suggests that laminin 1 does not mediate interaction with α6β4 at HDs. HD induction in vitro can be mediated by ECM preparations enriched in laminin 5. It therefore seems likely that it is laminin 5 rather than laminin 1 that acts as a ligand for α6β4 at the HD. It seems possible that laminin 1 may serve as a ligand for α6β4 in vivo in tissue where the basement membrane does not contain laminin 5. It is also probable that other ligand-receptor interactions occur at the plasma membrane to assemble and then stabilize the HD.

TYPE VII COLLAGEN AND THE SUB-LAMINA DENSA

Collagen VII is a disulphide-linked homotrimer of α1(VII) chains, and this single gene product is the major polypeptide component of the anchoring fibril (Fig. 6.3).[44,110] The human COL7A1 gene has been mapped to 3p21.3 and contains 118 exons separated by small introns and spanning about 31 kbp from transcription start site to polyadenylation site.[111,112] The subunit contains a long (424 nm) central triple helical domain[110] with one interruption of 39 aminoacids and several single interruptions at which glycine (in the GXY triplet sequence) is substituted by other residues. The position of the major interruption has been correlated with the position in the chain of a 'hinge' observed by rotary shadowing EM. The chain interruptions appear therefore to create a flexible helix. At the amino terminal end of the subunit is a large globular domain known as NC-1;[113] in this region the three subunits appear to be essentially independently folded, giving the appearance of a thin rod with three short, thicker branches at the end (Fig. 6.3).[110] The NC-1 domain contains several recognizable structural motifs:[113,114] at the extreme N-terminus is a domain resembling cartilage matrix protein, then there are nine consecutive fibronectin type III domains and one type A domain of the sort first identified in von Willebrand Factor. The type III and type A motifs are very widespread amongst ECM components and have been implicated in binding to other components of the ECM and cell surface as well as complement.[115,116] Within the type A domain is an RGD cell recognition motif. At the C-terminus of the α1(VII) subunit is a short non-helical domain (NC-2) that contains a Kunitz module.[44]

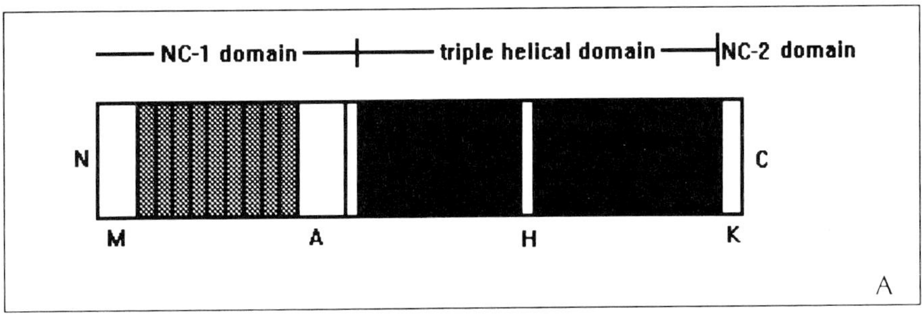

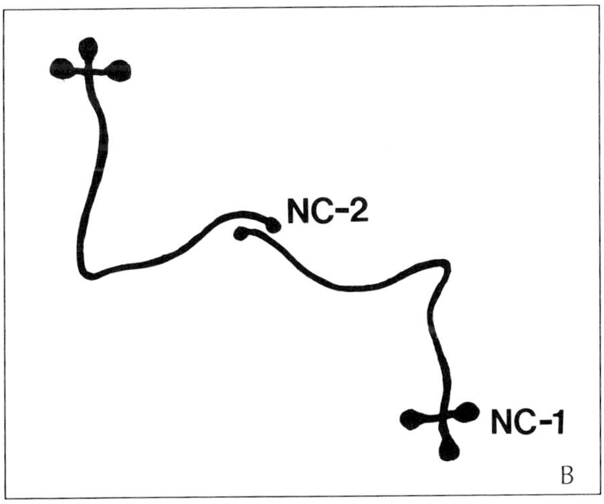

Fig. 6.3. (A) Type VII collagen polypeptide based on the structure of the cDNA. Adapted from reference 44. Regions of sequence similarity are identified as follows: M: cartilage matrix protein; A: type A domain as found in type VI collagen, von Willebrand Factor and numerous other ECM and complement components;[115] H: interruption in the triple helical sequence ('hinge'); K: Kunitz module. The grey stippled area represents nine successive fibronectin type III domains. (B) Anchoring fibrils are thought to contain disulphide-bonded dimers of type VII collagen in which interaction occurs between the NC-2 domain and a region in the triple helix. The NC-2 domains are removed after dimerization.

The collagen VII homotrimer self-associates by interactions at the C-terminal end of the molecule, possibly involving the binding of NC-2 to a region near the C-terminal end of the triple helix.[44,117] This produces a long fibril with NC-1 domains at both ends (Fig. 6.3).[44,110] The fibril has a characteristic centrosymmetric banding pattern. The NC-1 domain appears to be responsible for the interaction of collagen VII with components of the lamina densa including collagen IV,[114] and it may be that it also interacts with the basal epithelial cell surface. Such interactions could potentially be trivalent. Anchoring fibrils may loop from one area of basal lamina through the underlying stroma and back to an adjacent site. In this way fibrillar or microfibrillar (Fig. 6.1) components of the superficial stroma can be enclosed. This type of behavior indicates a linking function. Some anchoring fibrils, however, leave the lamina densa in an approximately orthogonal orientation and terminate deeper in the stroma. There they have been suggested to interact with 'anchoring plaques' which appear as islands of basal lamina isolated within the stroma.[44,47,118,119] In amnion, immunolocalization of laminin and type IV collagen has also been reported in the stroma and collagen VII may link the lamina densa to deeper structures in this tissue as well.[120] Antibodies to the NC-1 domain have been used to show that type VII collagen is found in tissues that produce HDs including skin, cornea, trachea, cervix, vagina and breast.[121] Furthermore, there is evidence that it originates in both epidermal and dermal cells.[32]

In cultured epidermal cell sheets grafted to burn sites it has been observed that there is a correlation between the appearance of collagen VII and HDs, both of which are present by day 11 after grafting.[122] At this stage the lamina densa was still discontinuous. When rabbit corneal epithelial sheets were separated from the underlying basement membrane, HD plaques were internalized by the cells. Recombination of these sheets with denuded cornea, the basement membrane of which was intact, led to the formation of new HDs within 1-3 days. A good correlation was observed between the new HDs and the sites of anchoring fibril insertion into the lamina densa.[123] In wounded rabbit cornea, primitive HDs are present after 66 hours healing in vivo, a stage at which the continuity of the lamina densa is not fully restored. Furthermore there is light microscopic immunolocalization evidence for the presence of BP230 as well as collagen VII.[124] Similarly, a correlation exists between the appearance of anchoring fibrils and HDs during corneal development.[124] When 13 day chick dermis and epidermis were separated and then recombined in the absence of a basal lamina, HDs with sub-basal dense plates and associated patches of lamina densa were present after only 2 days. After 3-4 days fragmentary basal lamina was present, and after 6-8 days the basal lamina became continuous and anchoring fibrils were formed.[125] However, no morphologically recognizable anchoring fibrils were shown and these data are no more than correlative. They are consistent with the role of anchoring fibrils in mechanical integration of the lamina densa and stroma in tissues bearing HDs.

Several cell types have been shown to produce respectable HDs in culture in the absence of morphologically recognizable anchoring fibrils. It seems unlikely that collagen VII is an absolute requirement for HD plaque formation if only because the established molecular dimensions and immunolocalization studies of the NC1 domain are inconsistent with direct contact between the anchoring fibril and the basal epithelial cell surface. However, it should be noted that collagen VII immunoreactivity may be present in the absence of morphologically distinct anchoring fibrils. It may be that it plays a role in the presentation of ligand (such as laminin 5) to the cell surface.

Direct evidence for the role of anchoring fibrils in preserving the mechanical integrity of the dermal-epidermal junction has come from the skin of individuals with the dystrophic form of epidermolysis bullosa,[44,52] which may be inherited in autosomal dominant or recessive forms or acquired (EBA). The disease varies in severity and in the skin area primarily affected. Electron microscopic studies first suggested that anchoring fibrils were absent, abnormal or diminished in number;[126,127] severe generalized recessive DEB skin lacks anchoring fibrils altogether, although immunoreactive collagen VII may still be detectable. In contrast, the HD plaques are usually of normal appearance.[52] Immunolocalization data indicate the absence of collagen VII from the subepithelial ECM,[128] though intracellular immunoreactivity has been reported in some patients, suggesting a block to secretion.[127,129,130] This suggests that collagen VII is not required to nucleate or stabilize HDs, but that it is required to stabilize the attachment of lamina densa to the subjacent ECM. Both dominant and recessive DEB have now been linked to the COL7A1 gene.[131,132] Furthermore, autoantibodies recognizing the NC1 domain have been detected in patients with acquired DEB.[133,134] These data link collagen VII convincingly to the DEB phenotype in which chronic blistering leads to severe scarring, fusion of the digits and loss of mobility in flexures.

It is beginning to be possible to correlate mutations occurring in COL7A1 with the disease phenotype. A single base substitution effecting a methionine to lysine conversion has been identified in a highly conserved region of the NC2 domain in a homozygotic patient with mild (mitis) DEB.[135] Subunit association during biosynthesis in the collagen family often begins at the C-terminus, and it is possible that this mutation affects chain association. Mutations giving rise to premature termination codons in the N-terminal domain, either in NC1 or the N-terminal triple helical region have also been reported.[57,136,137] These are associated with severe mutilating recessive DEB in homozygotes. Heterozygotes display variable density and morphology of anchoring fibrils. Affected members of a family showing dominant dystrophic DEB have been found to have a glycine to serine mutation in the triple helical region of collagen VII.[57] The mutation occurs within a sequence of 24 uninterrupted GXY repeats. This probably results in a less stable or conformationally altered anchoring fibril; electron microscopic examination indicated the presence of abnormally thin and more sparsely distributed anchoring fibrils.

SUMMARY

Disease studies indicate that laminin 5 (3 genes) and collagen VII (1 gene) are essential to the stability of the dermal-epidermal junction. Laminin 5 is a component of anchoring filaments and probably interacts with receptors in the epidermal basal cell surface at HDs. Collagen VII-containing anchoring fibrils function to join the lamina densa to underlying tissue, but are not an absolute requirement for the formation of HD plaques in adherent cells. The manner in which these components interact with each other, with other components of the basement membrane and with the cell surface remains to be elucidated.

ACKNOWLEDGMENT

We are grateful to Drs. Carolyn Jones and Farhad Behzad for the micrographs shown in Figure 6.1, and to Drs. J. Engel, G. Meneguzzi, J.P. Ortonne, K. Tryggvason and J. Uitto for access to unpublished data.

REFERENCES

1. Kelly DE. Fine structure of desmosomes, hemidesmosomes and an adepidermal globular layer in devéloping newt epidermis. J Cell Biol 1966; 28:51-72.
2. Eady RAJ. The basement membrane. Interface between the epithelium and the dermis. Structural features. Arch. Dermatol. 1988; 124:709-712.
3. Ghadially FN. Ultrastructural pathology of the cell and matrix. Volume II. London: Butterworths, 1988:1116.
4. Ellison J, Garrod DR. Anchoring filaments of the amphibian epidermal-dermal junction traverse the basal lamina entirely from

the plasma membrane of hemidesmosomes to the dermis. J Cell Sci 1984; 72:163-172.

5. Aplin JD, Campbell S, Allen TD. The extracellular matrix of human amniotic epithelium: ultrastructure, composition and deposition. J Cell Sci 1985; 79:119-136.
6. Legan PK, Collins JE, Garrod DR. The molecular biology of desmosomes and hemidesmosomes: What's in a name? Bio Essays 1992; 14:1-9.
7. Garrod DR. Desmosomes and hemidesmosomes. Curr Opinion Cell Biol 1993; 5:30-40.
8. Buck RC. Hemidesmosomes of normal and regenerating mouse corneal epithelium. Virchows Arch B Cell Pathol 1982; 41:1-16.
9. Gipson IK, Spurr-Michaud S, Tisdale A, Eelwell J, Stepp MA. Redistribution of the hemidesmosome components $\alpha6\beta4$ integrin and bullous pemphigoid antigens during epithelial wound healing. Exp Cell Res 1993; 207:86-98.
10. Tidman MJ, Eady RAJ. Ultrastructural morphometry of normal human dermal-epidermal junction. The influence of age, sex and body region on laminar and non-laminar components. J. Invest. Dermatol. 1984; 83:448-453.
11. Campbell S, Allen TD, Moser W, Aplin JD. The translaminal fibrils of the human amnion epithelium. J Cell Sci 1989; 94:307-318.
12. Aplin JD, Campbell S, Donnai P, Bard JBL, Allen TD. Importance of vitamin C in maintenance of the normal amnion: an experimental study. Placenta 1986; 7:337-389.
13. Hay ED. Development of the vertebrate cornea. Intl. rev. Cytol. 1980; 63:263-322.
14. Goldberg M, Escaig-Haye F. Is the lamina lucida of the basement membrane a fixation artefact? Eur J Cell Biol 1986; 42:365-368.
15. Chan FL, Inoue S, Leblond CP. The basement membranes of cryofixed or aldehyde-fixed, freeze-substituted tissues are composed of a lamina densa and do not contain a lamina lucida. Cell Tissue Res 1993; 273:41-52.
16. Yurchenko PD, O'Rear, J. In: Rohrbach DH, Timpl R, eds. Molecular and Cellular Aspects of Basement Membranes. Academic Press, 1993:19-47.
17. Kleppel MM, Michael AF. Expression of novel basement membrane components in the developing human kidney and eye. Am J Anat 1990; 187:165-174.
18. Mann K, Deutzmann R, Paulsson M, Timpl R. Characteriation of proteolytic fragments of the laminin-nidogen complex and their affinity in ligand binding assays. Eur J Biochem 1988; 178:71-80.
19. Aumailley M, Wiedemann H, Mann K, Timpl R. Binding of nidogen and the laminin-nidogen complex to basement membrane collagen type IV. Eur J Biochem 1989; 184:241-248.
20. Woodley DT, Rao CN, Hassell JR, Liotta LA, Martin GR, Kleinman HK. Interactions of basement membrane components. Biochim Biophys Acta 1983; 761:278-283.
21. Laurie GW, Bing JT, Kleinman HK, Hassell JR, Aumailley M, Martin GR, Feldmann RJ. Localization of binding sites for laminin, heparan sulfate proteoglycan and fibronectin on basement membrane (type IV) collagen. J Mol Biol 1986; 189:205-216.
22. Burgeson RE, Chiquet M, Deutzmann R, Ekblom P, Engel J, Kleinman H, Martin GR, Ortonne J-P, Paulsson M, Sanes J, Timpl R, Tryggvason K, Yamada Y, Yurchenco PD. A new nomenclature for laminins. Matrix Biol, 14:209-211.
23. Noonan DM, Hassell JR. Proteoglycans of basement membranes. In: Rohrbach DH, Timpl R, eds. Molecular and Cellular Aspects of Basement Membranes. Academic Press, 1993:189-210.
24. Battaglia C, Mayer U, Aumailley M, Timpl R. Basement membrane heparan sulphate proteoglycan binds to laminin by its heparan sulphate chains and to nidogen by sites in the protein core. Eur J Biochem 1992; 208:359-366.
25. Iozzo RV. Perlecan: a gem of a proteoglycan. Matrix Biol 1994; 14:203-208.
26. Timpl R, Aumailley M. Other basement membrane proteins and their calcium-binding potential. In: Rohrbach DH, Timpl R, eds. Molecular and Cellular Aspects of Basement Membranes. Academic Press, 1993:211-235.
27. Smith LT. Patterns of type VI collagen

compared to types I, III, and V collagen in human embryonic and fetal skin and in fetal skin-derived cultures. Matrix Biol 1994; 14:159-170.

28. Kielty CM, Shuttleworth CA. Synthesis and assembly of fibrillin by fibroblasts and smooth muscle cells. J Cell Sci 1993; 106:167-173.
29. Zimmermann DR, Dours-Zimmermann MT, Schubert M, Bruckner-Tuderman L. Versican is expressed in the proliferating zone in the epidermis and in association with the elastic network of the dermis. J Cell Biol 1994; 124:817-825.
30. Bosman FT, Cleutjens J, Beek C, Havenith M. Basement membrane heterogeneity. Histochem J 1989; 21:629-633.
31. Inoue S. Ultrastructure of basement membranes. Int Rev Cytol 1989; 117:57-98.
32. Marinkovich MP, Keene, DR, Rimberg CS, Burgeson RE. Cellular origin of the dermal-epidermal junction. Dev. Dynamics 1993; 197:255-267.
33. Owaribe K, Kartenbeck J, Stumpp S, Magin TM, Krieg T, Diaz LA, Franke WW. The hemidesmosomal plaque. 1. Characterization of a major constituent protein as a differentiation marker for certain forms of epithelia. Differentiation 1990; 45:207-220.
34. Sawamura D, Li K, Chu M-L, Uitto J. Human bullous pemphigoid antigen (BPAG1). Amino acid sequences deduced from cloned cDNAs predict biologically important peptide segments and protein domains. J Biol Chem 1991; 266: 17784-17790.
35. Tamai K, Sawamura D, Choi HY, Li K, Uitto J. Molecular biology of the 230kDa bullous pemphigoid antigen. Cloning of the BPAG1 gene and its tissue-specific expression. Dermatol 1994; 189:27-33.
36. Hiede Y, Nishizawa Y, Uematsu J, Owaribe K. Identification of a new hemidesmosomal protein, HD1: a major, high molecular mass component of isolated hemidesmosomes. J Cell Biol 1992; 116: 1497-1506.
37. Stepp MA, Spurr-Michaud S, Tisdale A, Elwell J, Gipson IK. α6β4 integrin heterodimer is a component of hemidesmosomes. Proc Nat Acad Sci USA 1990; 87:8970-8974.
38. Sonnenberg A, Calafat J, Hanssen H, Daams H, Van Der Raaij-Helmer LMH, Falcioni R, Kennel SJ, Aplin JD, Baker J, Loizidou M, Garrod D. Integrin α6β4 complex is located in hemidesmosomes, suggesting a major role in epidermal cell-basement membrane adhesion. J Cell Biol 1991; 113:907-917.
39. Behzad F, Jones CJP, Aplin JD. The role of integrin α6β4 in hemidesmomes of human amnion. Biochem Soc Trans 1991; 19:381S.
40. Li K, Giudice GJ, Tamai K, Do HC, Sawamura D, Diaz LA, Uitto J. Cloning of partial cDNA for mouse 180kDa bullous pemphigoid antigen (BPAG2), a highly conserved collagenous protein of the cutaneous basement membrane zone. J Invest Dermatol 1992; 260: 258-263.
41. Giudice GJ, Siquiera HL, Elias PM, Diaz LA. Identification of two collagen domains within the bullous pemphigoid autoantigen, BP180. J Clin Invest 1991; 87:734-738.
42. Giudice GJ, Emery DJ, Diaz LA. Cloning and primary structural analysis of the bullous pemphigoid autoantigen BP180. J Invest Dermatol 1992; 99:243-250.
43. Rousselle P, Lunstrum GP, Keene DR, Burgeson RE. Kalinin: an epithelium-specific basement membrane adhesion molecule that is a component of anchoring filaments. J Cell Biol 1991; 114:567-576.
44. Burgeson RE. Type VII collagen, anchoring fibrils and epidermolysis bullosa. J Invest Dermatol 1993; 101:252-255.
45. Fine JD, Horiguchi Y, Jester J, Couchman JR. Detection and characterisation of a midlamina lucida hemidesmosome-associated antigen (19-DEJ-1) present within human skin. J Invest Dermatol 1989; 92:825-830.
46. Klatte DH, Kurpakus MA, Grelling KA, Jones JCR. Immunochemical characterization of three components of the hemidesmosome and their exprssion incultured epithelial cells. J. Cell Biol. 1989; 109:377-3390.
47. Clermont Y, Hernot L. Structure of the complex basement membrane underlying the epithelium of the vas deferens in the rat. Anat Rec 1988; 221:482-493.
48. Sonnenberg A, Linders CJT, Daams JH, Kennel SJ. The α6β1 (VLA-6) and α6β4 protein complexes: tissue distribution and

biochemical properties. J Cell Sci 1990; 96:207-217.

49. Kennel SJ, Godfrey V, Chang LY, Lankford TK, Foote LJ, Makkinje A. The β4 subunit of the integrin family is displayed on a restricted subset of endothelium in mice. J Cell Sci 1992; 101:145-150.
50. Phillips RJ, Aplin JD, Lake BD. Antigenic expression of integrin α6β4 in junctional epidermolysis bullosa. Histopathology 1994; 24:571-576.
51. Aplin JD. Expression of integrin α6β4 in human trophoblast and its loss from extravillous cells. Placenta 1993; 14:203-215.
52. Bruckner-Tuderman L. Epidermolysis bullosa. In: Lapierre CM, Krieg T, eds. Connective Tissue Diseases of the Skin. New York: Marcel Dekker, 1993:263-281.
53. Timpl R, Rohde H, Gehron Robey. P, Rennard SI, Foidart JM, Martin GR. Laminin—A glycoprotein from basement membranes. J Biol Chem 1979; 254:9933-9937.
54. Engel J. Laminins and other strange proteins. Biochemistry 1992; 31:10643-10651.
55. Engvall E. Laminin variants: why, where and when? Kidney Int 1993;43:2-6.
56. Engel, J. Structure and function of Laminin. In: Rhorbach DH, Timpl R, eds. Molecular and Cellular Aspects of Basement Membranes. Academic Press, 1993:147-176.
57. Uitto J, Pulkkinen L, Christiano AM. Molecular basis of the dystrophic and junctional forms of epidermolysis bullosa: mutations in the type VII collagen and kalinin (laminin 5) genes. J Invest Dermatol, in press.
58. Leivo I, Engvall E. Merosin a protein specific for basement membranes of schwann cells, striated muscle, and trophoblast is expressed late in nerve and muscle development. Proc Natl Acad Sci USA 1988; 85:1544-1548.
59. Paulsson M, Saladin K. Mouse heart laminin. Purification of the native protein and structural comparison with Engelbreth-Holm-Swarm tumour laminin. J Biol Chem 1989; 264:18726-18732.
60. Paulsson M, Saladin K, Engvall E. Structure of laminin variants. The 300-kDa chains of murine and bovine heart laminin are related to the human placenta merosin heavy chain and replace the A chain in some laminin variants. J Biol Chem 1991; 266:17545-17551.
61. Hunter DD, Shah V, Merlie JP, Sanes JR. A laminin-like adhesive protein concentrated in the synaptic cleft of the neuromuscular junction. Nature 1989; 338:229-233.
62. Engvall E, Earwicker D, Haaparanta T, Ruoslahti E, Sanes JR. Distribution and isolation of four laminin variants; tissue restricted distribution of heterotrimers assembled from five different subunits. Cell Reg 1990; 1:731-740.
63. Paulsson M. Laminin and collagen IV variants and heterogeneity in basement membrane composition. In: Rohrbach DH, Timpl R, eds. Molecular and Cellular Aspects of Basement Membranes. Academic Press, 1993:177-188.
64. Verrando P, Hsi, B-L, Yeh, C-J, Pisani A, Serieys N, Ortonne J-P. (1987) Monoclonal antibody GB3, a new probe for the study of human basement membranes and hemidesmosomes. Exp Cell Res 1987; 170:116-128.
65. Burgeson R. Dermal-epidermal adhsion in skin. In: Rohrbach DH, Timpl R, eds. Molecular and Cellular Aspects of Basement Membranes. Academic Press, 1993:49-66.
66. Paulsson M, Aumailley M, Deutzmann R, Timpl R, Beck K, Engel J. Laminin-nidogen complex: extraction with chelating agents and structural characterisation. Eur J Biochem 1987; 166:11-19.
67. Mecham RP. Receptors for laminin in mammalian cells. FASEB J 1991; 5:2538-2546.
68. Gehlsen KR, Dickerson K, Argraves WS, Engvall E, Ruoslahti E. Subunit structure of a laminin-binding integrin and localization of its binding site on laminin. J Biol Chem 1989; 264:19034-19038.
69. Hall DE, Reichardt LF, Crowley E, Holley B, Moezzi H, Sonnenberg A, Damsky CH. The α1β1 and α6β1 integrin heterodimers mediate cell attachment to distinct sites on laminin. J Cell Biol 1990; 110:2175-2184.
70. Hunter DD, Porter BE, Bulock JW, Adams AP, Merlie JP, Sanes JP. Primary sequence of a motor neuron-selective adhesive site in the synaptic basal lamina protein s-laminin. Cell 1989; 59: 905-913.
71. Marinkovich MP, Lunstrum P, Burgeson RE. The anchoring filament protein kalinin

is synthesized and secreted as a high molecular weight precursor. J Biol Chem 1992; 267:17900-17906.

72. Verrando P, Pisani A, Ortonne J-P. The new basement membrane antigen recognized by the monoclonal antibody GB3 is a large size glycoprotein: modulation of its expression by retinoic acid. Biochim Biophys Acta 1988; 942:45-56.
73. Marinkovich MP, Verrando P, Keene DR, Meneguzzi GP, Lunstrum GP, Ortonne JP, Burgeson RE. The basement membrane proteins kalinin and nicein are structurally and immunologically identical. Lab Invest 1993; 69:295-299.
74. Gereke D, Wagman DW, McDonough AM, Burgeson RE. cDNAs encoding the three chains of the anchoring filament protein kalinin show similarity to the laminin A, B1, and B2 chains. Mol Biol Cell 1992; 3:1a (Abs 2).
75. Vailly J, Szepetowski P, Mattei MG, Pedeutour F, Burgeson R, Ortonne JP, Meneguzzi G. The genes for nicein/kalinin 125k- and 100-kDa subunits, candidates for junctional epidermolysis bullosa, map to chromosomes 1q32 and 1q25-q31. Genomics; 1994; 21:286-288.
76. Aberdam D, Galliano MF, Mattei MG, Pisani-Spadafora A, Ortonne JP, Meneguzzi G. Assignment of mouse nicein genes to chromosomes 1 and 18. Mammalian Genome; 1994; 5:229-233.
77. Aberdam D, Aguzzi A, Baudoin C, Galliano MF, Ortonne J-P, Meneguzzi G. Developmental expression of nicein adhesion protein (laminin-5) subunits suggests multiple morphogenic roles. Cell Adhesion and Communication, 1994; 2:115-129.
78. Baudoin C, Galliano MF, Verrando P, Ortonne JP, Meneguzzi G. The 150 kD subunit of the basement membrane component BM600/ nicein shares significative homology with the A chain of human laminin. J Invest Dermatol; in press.
79. Gerecke DR, Wagman DW, Champliaud MF, Burgeson RE. The complete primary structure for a novel laminin chain the laminin B1k chain. J Biol Chem, 1994; 269: 11073-11080.
80. Vailly J, Verrando P, Champliaud M-F, Gerecke D, Wagman DW, Baudoin C, Aberdam D, Burgeson R, Bauer E, Ortonne J-P. The 100-kDa chain of nicein/kalinin is a laminin B2 chain variant. Eur J Biochem 1994; 219:209-218.
81. Kalluuki P, Sainio K, Eddy R, Byers M, Kalluuki T, Sariola H, Beck C, Hirvonen H, Shows T, Tryggvason K. A truncated laminin chain homologous to the B2 chain: structure, spatial expression, and chromosomal assignment. J Cell Biol 1992; 119:679-693.
82. Heagarty AHM, Eady RAJ, Kennedy AR, Nicolaides KH, Rodeck CH, Hsi BL, Ortonne JP. Rapid prenatal diagnosis of epidermolysis bullosa letalis using GB3 monoclonal antibody. Br J Dermatol 1987; 117:271-276.
83. Smith LT, Sakai LY, Burgeson RE, Holbrook KA. Ontogeny of structural components at the dermal-epidermal junction in human embryonic and fetal skin: the appearance of anchoring fibrils and type VII collagen. J Invest Dermatol 1988; 90:480-485.
84. Scott JH, Symons NBB. Introduction to dental anatomy, 8th edition. Churchill Livingstone, Edinburgh. 1977; p189.
85. Marinkovich MP, Lunstrum GP, Keene DR, Burgeson RE. The dermal-epidermal junction of human skin contains a novel laminin variant. J Cell Biol 1992; 119:695-703.
86. Carter WG, Ryan MC, Gahr PJ. Epiligrin, a new cell adhesion ligand for integrin $\alpha3\beta1$ in epithelial basement membranes. Cell 1991; 65:599-610.
87. Domloge-Hultsch N, Gammon WR, Briggaman RA, Gil SG, Carter WG, Yancey LB. Epiligrin, the major human keratinocyte integrin ligand, is a target in both an acquired autoimmune and an inherited subepidermal blistering skin disease. J Clin Invest 1992; 90:1628-1633.
88. Tidman MJ, Eady RAJ. Hemidesmosome heterogeneity in junctional epidermolysis bullosa revealed by morphometric analysis. J Invest Dermatol 1986; 86:51-56.
89. Meneguzzi G, Marinkovich MP, Verrando P, Vailly J, Pisani A, Burgeson R, Ortonne JP. Abnormal expression of kalinin in lethal junctional epidermolysis bullosa. J In-

vest Dermatol 1992; 98:590 (A).

90. Verrando P, Blanchet-Bardon C, Pisani A, Thomas L, Cambazard F, Eady RAJ, Schofield O, Ortonne J-P. Monoclonal antibody GB3 defines a widespread defect of several basement membranes and a keratinocyte dysfunction in patients with lethal junctional epidermolysis bullosa. Lab Invest 1991; 64:85-92.
91. Fine JD, Bauer EA, Briggman RA, Carter DM. Revised clinical and laboratory criteria for sub-types of inherited epidermolysis bullosa. J Am Acad Dermatol 1991; 24:119-135.
92. Hegarty AHM, Kennedy AR, Eady RAJ, Hsi BL, Verrando P, Yeh CJ, Ortonne JP. GB3 Monoclonal antibody for diagnosis of junctional epidermolysis bullosa. Lancet April 1986; 860.
93. Verrando P, Schofield O, Ishida-Yamamoto A, Aberdam D, Partouche O, Eady RAJ, Ortonne JP. Nicein (BM600) in junctional epidermolysis bullosa: polyclonal antibodies provide new clues for pathogenic role. J Invest Dermatol 1993; 101:738-743.
94. Baudoin C, Miquel C, Blanchet-Bardon C, Gambini C, Meneguzzi G, Ortonne JP. Herlitz junctional epidermolysis bullosa keratinocytes display heterogeneous defects of nicein/kalinin gene expression. J Clin Invest 1994; 93:862-869.
95. Aberdam D, Galliano M-F, Vailly J, Pulkkinen L, Bonifas J, Christiano AM, Tryggvason K, Uitto J, Epstein EH, Ortonne J-P, Meneguzzi G. Herlitz's junctional epidermolysis bullosa is linked to mutations in the gene (LAMC2) for the γ2 subunit of nicein/kalinin (LAMININ-5). Nature Genetics 1994; 6:299-304.
96. Pulkkinen L, Christiano AM, Airenne T, Haakana H, Tryggvason K, Uitto J. Mutations in the γ2 chain gene (LAMC2) of kalinin/laminin 5 in the junctional forms of epidermolysis bullosa. Nature Genetics 1994; 6:293-297.
97. Riddelle KS, Green KJ, Jones JCR. Formation of hemidesmosomes in vitro by a transformed rat bladder cell line. J Cell Biol 1991; 112:159-168.
98. Jones JCR, Kurpakus MA, Cooper HM, Quaranta V. A function for the integrin α6β4 in the hemidesmosome. Cell Reg 1991; 2, 427-438.
99. Langhofer M, Hopkinson SB, Jones JCR. The matrix secreted by 804G cells contains laminin-related components that participate in hemidesmosome assembly in vitro. J Cell Sci 1993; 105:753-764.
100. Sonnenberg A, de Melker AA, Martinez de Velasco AM, Janssen H, Calafat J, Niessen CM. Formation of hemidesmosomes in cells of a transformed murine mammary tumour cell line and mechanisms involved in adherence of these cells to laminin and kalinin. J Cell Sci 1993; 106:1083-1102.
101. Kurpakus MA, Quaranta V, Jones JCR. Surface relocation of α6β4 intergrins and assembly of hemidesmosomes in an in vitro model. J Cell Biol 1991; 115:1737-1750.
102. Hynes RO. Integrins: versatility, modulation, and signaling in cell adhesion. Cell 1992; 69:11-25.
103. Rousselle P, Aumailley M. Kalinin is more efficient than laminin in promoting adhesion of primary keratinocytes and some other epithelial cells and has a different requirement for integrin receptors. J Cell Biol 1994; 125:205-214.
104. Niessen CM, Horgervorst F, Jaspers LH, De Melker AA, Delwel GO, Hulsman EHM, Kuikman I, Sonnenberg A. The α6β4 integrin in a receptor for both laminin and kalinin. Exp Cell Res 1994; 211: 360-367.
105. Lee EC, Lotz MM, Steele GD, Mercurio AM. The integrin α6β4 is a laminin receptor. J Cell Biol 1992; 117, 671-678.
106. Aplin JD, Sattar A, Mould P. Variant choriocarcinoma (BeWo) cells that differ in adhesion and migration on fibronectin display conserved patterns of integrin expression. J Cell Sci 1992; 103:435-444.
107. Sonnenberg A, Linders CJT, Modderman PW, Damsky CH, Aumailley M, Timpl R. Integrin recognition of different cell-binding fragments of laminin (P1, E3, E8) and evidence that α6β1 but not α6β4 functions as a major receptor for fragment E8. J Cell Biol 1990; 110:2145-2155.
108. Delwel GO, Hogervorst F, Kuikman I, Paulsson M, Timpl R, Sonnenberg A. Expression and function of the cytoplasmic variant of the integrin α6 subunit in trans-

fected K562 cells: activation-dependent adhesion and interaction with isoforms of laminin. J Biol Chem 1993; 268:25865-25875.

109. Marchisio PC, Cremona O, Savoia P, Pellegrini G, Ortonne J-P, Verrando P, Burgeson RE, Cancedda R, De Luca M. The basement membrane protein BM-600/nicein codistributes with kalinin and the integrin α6β4 in human cultured keratinocytes. Exp Cell Res 1993; 205:205-212.

110. Bächinger HP, Morris NP, Lunstrum GP, Keene DR, Rosenbaum LM, Compton LA, Burgeson RE. The relationship of the biophysical and biochemical characteristics of type VII collagen to the function of anchoring fibrils. J Biol Chem 1990; 265:10095-10101.

111. Parente MG, Chung LC, Ryynänen J, Woodley DT, Wynn KC, Bauer EA, Mattei M-G, Uitto J. Human type VII collagen: cDNA cloning and chromosomal mapping of the gene. Proc Nat Acad Sci USA 1991; 88:6931-6935.

112. Christiano AM, Hoffman GG, Chung-Honet LC, Lee S, Cheng W, Uitto J, Greenspan D. Structural organisation of the human type VII collagen gene (COL7A1), composed of more exons than any previously characterised gene. Genomics 1994; 21:169-179.

113. Christiano AM, Rosenbaum LR, Chung-Honet LC, Parente MG, Woodley DT, Pan TC, Zhang RZ, Chu ML, Burgeson RE, Uitto J. The large non-collagenous domain NC-1 of type VII collagen is amino terminal and chimeric. Homology to cartilage matrix protein, the type III domain of fibronectin and the A domains of von Willebrand factor. Human Mol Genet 1992; 7:475-481.

114. Gammon WR, Abernethy ML, Padilla KM, Prisayanh PS, Cook ME, Wright J, Briggaman RA, Hunt SW. Noncollagenous (NC1) comain of collagen VII resembles multidomain adhesion proteins involved in tissue-specific organization of extracellular matrix. J Invest Dermatol 1992; 99:691-696.

115. Colombatti A, Bonaldo P, Doliana R. Type A modules: interacting domains found in several non-fibrillar collagens and in other extracellular matrix proteins. Matrix 1993; 13:297-306.

116. Dickinson CD, Veerapandian B, Dai XP, Hamlin RC, Xuong NH, Ruoslahti E, Ely KR. Crystal structure of the tenth type III cell adhesion module of human fibronectin. J Mol Biol 1994; 236:1079-1092.

117. Morris NP, Keene DR, Glanville RW, Bentz H, Burgeson RE. The tissue form of type VII collagen is an antiparallel dimer. J Biol Chem 1986; 261:5638-5644.

118. Keene DR, Sakai LY, Morris NP, Burgeson RE. Type VII collagen forms an extended network of anchoring fibrils. J Cell Biol 1987; 104:611-621.

119. Gipson IK, Spurr Michaus SJ, Tisdale AS. Anchoring fibrils form a complex network in human and rabbit cornea. Invest Opthal Vis Sci 1986; 28:212-220.

120. Ockleford C, Bright N, Hubbard A, D'Lacey C, Smith J, Gardiner L, Sheikh T, Albentosa M, Turtle K. Micro-trabeculae, macroplaques or mini-basement membranes in human term fetal membranes? Phil Trans R Soc Lond B 1993; 342:121-136.

121. Wetzels RHW, Robben HCM, Leigh IM, Schaafsma HE, Vooijs GP, Ramaekers FCS. Distribution patterns of type VII collagen in normal and malignant human tissues. Am J Pathol 1991; 139:451-459.

122. Mommaas AM, Teepe RGC, Leigh IM, Mulder AA, Koebrugge EJ, Vermeer BJ. Ontogenesis of the basement membrane zone after grafting cultured human epithelium: a morphologic and immunoelectron microscopic study. J Invest Dermatol 1992; 99:71-77.

123. Gipson IK, Grill SM, Spurr SJ, Brennan SL. Hemidesmosome formation in vitro. J. Cell Biol. 1983; 97:849-857.

124. Gipson IK, Spurr-Michaud SJ, Tisdale AS. Hemidesmosomes and anchoring fibril collagen appear synchronously during development and wound healing. Dev Biol 1988; 126:253-262.

125. Akimoto Y, Obinata A, Endo H, Hirano H. Reconstruction of basement membrane in recombinants of epidermis and dermis of chick embryonic skin in vitro: an electron microscopic study. Anat Rec 1991; 231:375-382.

126. Bruckner-Tuderman L, Rantala I, Reunala

T. Evidence for a structural abnormality of collagen VII in a patient with dystrophic epidermolysis bullosa inversa. J Invest Dermatol 1992; 98:141-146.

127. McGrath JA, Ishida-Yamamoto A, O'Grady A, Leigh IM, Eady RA. Structural variations in anchoring fibrils in dystrophic epidermolysis bullosa: correlation with type VII collagen expression. J Invest Dermatol 1993; 100:366-372.

128. Bruckner-Tuderman L, Guscetti F, Ehrensperger F. Animal model for dermolytic mechanobullous disease: sheep with recessive dystrophic epidermolysis bullosa lack collagen VII. J Invest Dermatol 1991; 96:452-458.

129. Fine J-D, Johnson LB, Cronce D, Wright JY, Leigh IM, McCollough M, Briggaman RA. Intracytoplasmic retention of type VII collagen and dominant dystrophic epidermolysis bullosa: reversal of defect following cessation of or marked improvement in disease activity. J Invest Dermatol 1993; 101:232-236.

130. Phillips RJ, Harper JI, Lake BD. Intraepidermal collagen type VII in dystrophic epidermolysis bullosa: report of five new cases. Br J Dermatol 1992; 126:222-230.

131. Ryynänen M, Ryynänen J, Sollberg S, Iozzo RV, Knowlton RG, Uitto J. Genetic linkage of type VII collagen (COL7A1) to dominant dystrophic epidermolysis bullosa in families with abnormal anchoring fibrils. J Clin Invest 1992; 89:974-980.

132. Hovnanian A, Duquesnoy P, Blanchet-Bardon C, Knowlton RG, Amselem S, Lathrop M, Dubertret L, Uitto J, Goossens M. Genetic linkage of recessive dystrophic epidermolysis bullosa to the type VII collagen gene. J Clin Invest 1992; 90:10321036.

133. Shimizu H, McDonald JN, Gunner DB, Black MM, Bhogal B, Leigh IM, Whitehead PC, Rady RAJ. Epidermolysis bullosa acquisita antigen and the carboxyl terminus of type VII collagen have a common immunolocalization to anchoring fibrils and lamina densa of basement membrane. Br J Dermatol 1990; 122:577-585.

134. Lapiere JC, Woodley DT, Parente MG, Iwasaki T, Wynn KC, Christiano AM, Uitto J. Epitope mapping of type VII collagen. Identification of discrete peptide sequences recognized by sera from patients with acquired epidermolysis bullosa. J Clin Invest 1993; 92:1831-1839.

135. Christiano AM, Greenspan DS, Hoffman GG, Zhang X, Tamai Y, Lin AN, Dietz HC, Hovnanian A, Uitto J. A missense mutation in type VII collagen in two affected siblings with recessive dystrophic epidermolysis bullosa. Nat Genet 1993; 4:62-66.

136. Hilal L, Rochat A, Duquesnoy P, Blanchet-Bardon C, Wechsler J, Martin N, Christiano AM, Barrandon Y, Uitto J, Goossens M. A homozygous insertion-deletion in the type VII collagen gene (COL7A1) in Hallopeau-Siemens dystrophic epidermolysis bullosa. Nat Genet 1993; 5:287-293.

137. Christiano AM, Anhalt G, Gibbons S, Bauer EA, Uitto J. Premature termination codons in the type VII collagen gene (COL7A1) underlie severe mutilating recessive dystrophic epidermolysis bullosa. Genomics 1994; 21:160-168.

CHAPTER 7

Desmosome, Hemidesmosome and Disease

John R. Stanley, Sarolta Kárpáti

What do the discovery of insulin and that of proto-oncogenes have in common? Both illustrate that we often don't appreciate how something works until it goes wrong. The discovery of insulin hinged on the understanding of diabetes while that of proto-oncogenes relied on the study of viral carcinogenesis. Or put another way, diseases hold many clues for our understanding of the normal structure and function of tissues and organs.

The skin is no exception to this general rule. In this chapter I will discuss how pemphigus and bullous pemphigoid, autoantibody-mediated blistering skin diseases, illustrate the crucial importance of cell adhesion in epidermis and define molecules involved in that function.

WHAT'S WRONG WITH THE PATIENTS?

There are two types of pemphigus, pemphigus vulgaris (PV) and pemphigus foliaceus (PF).[1] Patients with both types of pemphigus develop blisters. Histology of these blisters shows that they occur because the keratinocytes in the epidermis lose their ability to adhere to each other. The only difference in histology is that in PV the loss of adhesion occurs deep in the epidermis, whereas in PF the pathology is more superficial (Fig. 7.1).

The prognosis of these diseases underscores the importance of keratinocyte cell adhesion in maintaining the normal function of epidermis. If PV is left untreated, the patient will almost certainly die from failure of skin function.[2] Over the course of several months to years after diagnosis, patients with PV experience increasing involvement of mucous membranes and skin. Mucous membrane blisters are extremely fragile and quickly evolve into painful erosions. In the mouth and pharynx these lesions prevent adequate nutrition and fluid intake. Flaccid blisters on the skin evolve into large areas of erosions which are sites of fluid loss

Molecular Biology of Desmosomes and Hemidesmosomes, edited by Jane E. Collins and David R. Garrod.

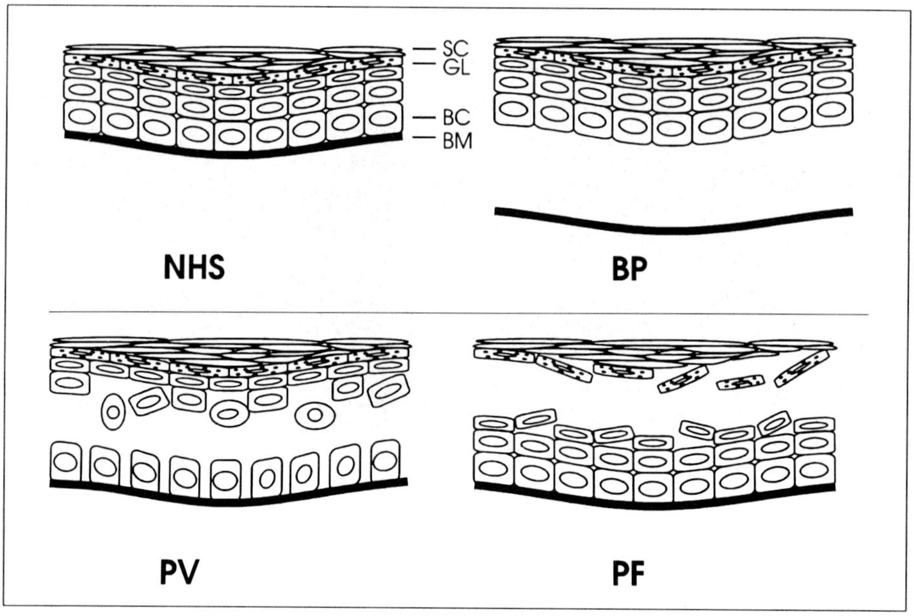

Fig. 7.1 Schematic cross-section of epidermis showing the blister localization in bullous pemphigoid (BP), pemphigus vulgaris (PV), and pemphigus foliaceus (PF). NHS, normal human skin; SC, stratum corneum; GL, granular layer; BC, basal cells; BM, basement membrane. Reprinted with permission from Stanley JR, Adv Immmunol 1993; 53:291-325.

and potential infection. Within 5 years almost all untreated patients will die, usually from infection.

Because the blister in PF is more superficial in the epidermis than is that in PV, the blister roof is more fragile and it is difficult to detect intact blisters in these patients because they rapidly evolve into crusted erosions. In addition, for unknown reasons, PF patients do not develop mucous membrane lesions. Thus, these patients usually manifest crusted, scaly erosions on the skin. Because more of the underlying epidermis is intact than in PV erosions, there is less propensity for fluid loss and infection in PF. However, even though the prognosis of PF is better, there is still a significant mortality in untreated disease.[2]

Bullous pemphigoid (BP), as the name implies, is also a blistering disease.[2] However, unlike pemphigus, the blister results from loss of keratinocyte cell adhesion to the underlying basement membrane (Fig. 7.1) and cell-to-cell adhesion is maintained intact. Because the blister roof is composed of the entire epidermis, it is not as fragile as that of pemphigus patients. Thus, BP patients characteristically develop tense skin blisters. If the blisters involve extensive areas of the skin surface, BP may be fatal due to loss of the normal function of the skin as a barrier against fluid loss and infection.

WHAT CAUSES THE DISEASES?

Pemphigus and BP patients have autoantibodies in their skin and serum.[3,4] These autoantibodies bind to the sites of pathology in these diseases. That is, BP autoantibodies bind to the basement membrane of stratified squamous epithelia whereas pemphigus autoantibodies (of both types) bind to the cell surface of the same tissue (Fig. 7.2).

In pemphigus of both types these autoantibodies cause the blisters.[1] This conclusion is based on clinical and experimental data. Disease activity usually correlates with amount of autoantibodies in the serum, as measured by the indirect immunofluorescence titer.[5,6] In passive transfer

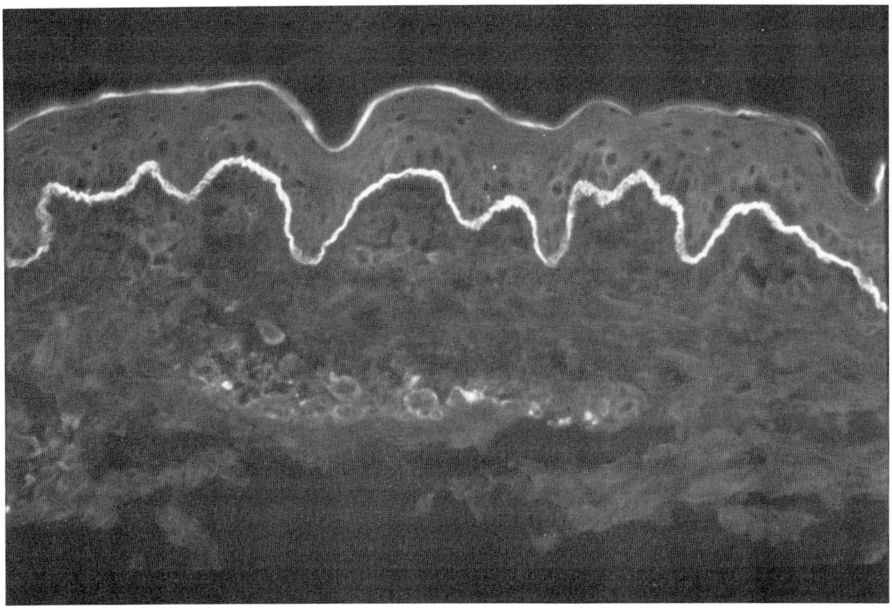

A

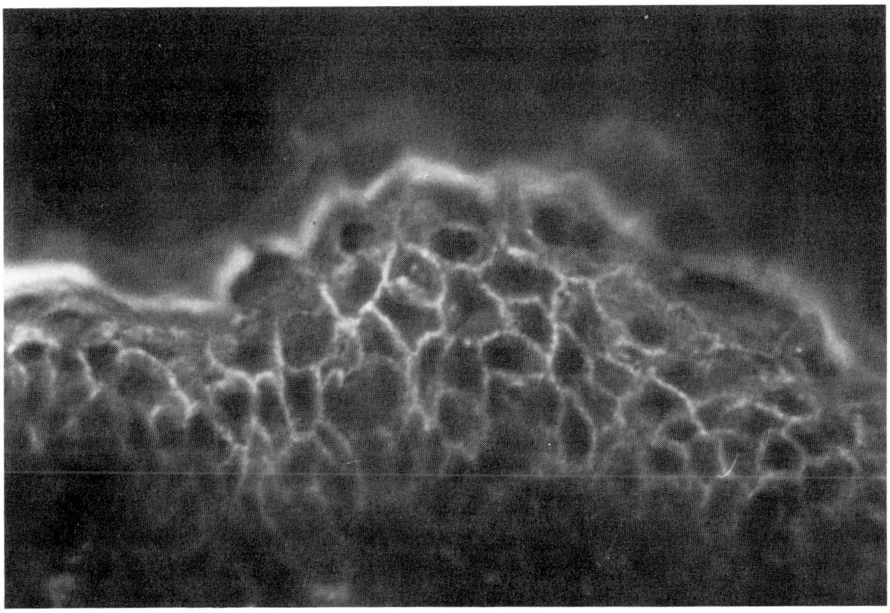

B

Fig. 7.2 Indirect immunofluorescence with BP and pemphigus sera. (A) BP IgG localizes to the epidermal basement membrane zone of pig skin. (There is also non-specific fluorescence of the stratum corneum). (Magnification x 215) (B) PF IgG localizes to the cell surface of keratinocytes in normal human skin. (There is also non-specific fluorescence in the lower stratum corneum). (Magnification x 565).

experiments of nature, transplacental passage of autoantibodies causes transient disease in neonates of mothers with PV.[7] However, perhaps because the antibodies are not as efficiently transferred, PF mothers do not have affected babies.[8]

Experimentally, blisters can be induced by adding PV or PF IgG to cultured skin (Fig. 7.3).[9,10] These antibodies cause loss of epidermal cell adhesion without the involvement of inflammatory cells or complement, implying a rather direct effect of the autoantibodies in causing pathology. Finally, both PV and PF can be induced in neonatal mice by passive transfer of human autoantibodies (Fig. 7.4).[11,12] Even in the absence of complement these mice will develop disease.[13] Interestingly even monovalent Fab' fragments of PF IgG will induce disease, demonstrating that cross-linking of cell surface molecules is probably not involved in pathogenesis.

These clinical and experimental observations demonstrate that autoantibodies in PV and PF are pathogenic and do not require additional inflammatory mechanisms to mediate their damage.

The involvement of BP autoantibodies in pathogenesis is not as straightforward. Whereas lesions of pemphigus may show few, if any, inflammatory cells, those of BP almost always have some neutrophils and eosinophils.[2] These inflammatory cells are probably recruited by the activation of complement by BP IgG.[14-17] Consistent with this idea, direct immunofluorescence of BP patients' skin always shows C3 (the third component of complement) fixed (by IgG) at the basement membrane[18] Thus, a contributing factor to the lesion production in BP may be protease release from inflammatory cells recruited to the basement membrane from activation of complement by autoantibodies.[17]

BP IgG, alone, added to skin organ culture does bind the basement membrane, yet does not cause subepidermal blisters, perhaps because complement and inflammatory cells are absent. It is surprising, however, that BP IgG isolated from BP sera does not cause disease in neonatal mice since they would be able to supply complement and inflammatory cells. However, just recently, it has been shown that the

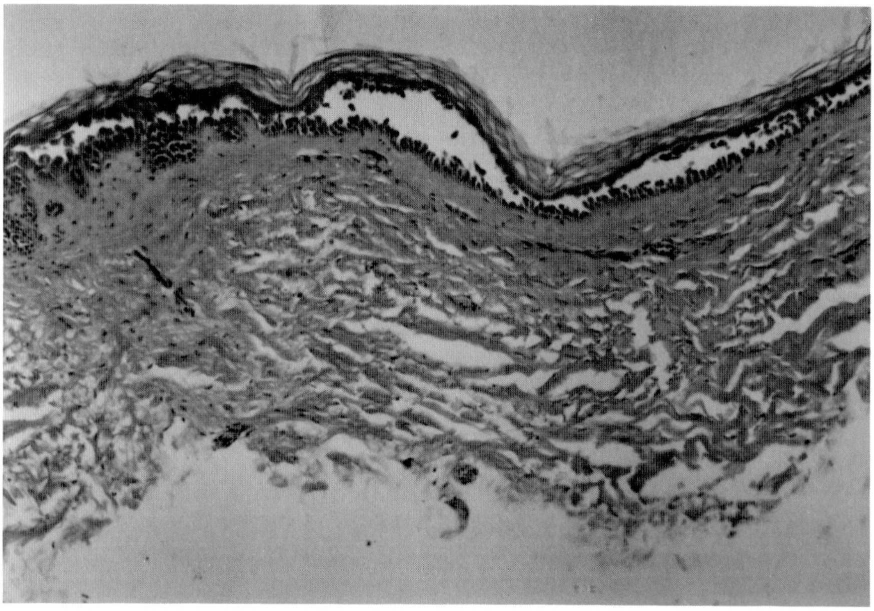

Fig. 7.3 Organ culture of normal human skin incubated with PV serum for 48 hours shows suprabasilar loss of cell-to-cell adhesion in the epidermis. (Magnification x 110).

reason for this finding may be that mice do not express a major epitope that antibodies from most BP patients bind on one of the BP antigens (called BPAG2).[19] Rabbit antibodies raised against the corresponding mouse epitope do induce subepidermal blisters in neonatal mice, strongly arguing for a role of BP autoantibodies in causing disease.[19] The lesional skin in these mice did tend to show C3 and inflammatory cells at the epidermal basement membrane. However, it was also postulated that antibodies to this specific epitope might directly interfere with keratinocyte-basement membrane adhesion.

WHAT DO THE AUTOANTIBODIES TELL US ABOUT CELL ADHESION MOLECULES?

Because pemphigus and BP are autoantibody-mediated diseases due to loss of keratinocyte cell adhesion it was perhaps reasonable, although maybe naive, to assume that the antibodies from these patients would define new or identify previously known cell adhesion molecules.

This assumption was first confirmed for PF antigen. Immunoblotting and immunoprecipitation studies defined PF antigen as a 160 kD glycoprotein that, within epidermis, was complexed, partially through disulfide binding, in a stoichiometric ratio, with an 85 kD protein.[20-22] Interestingly, antibodies from most PF patients bound a calcium-sensitive conformational epitope on the 160 kD PF antigen.[21] The PF antigen was soon shown to be identical to desmoglein, the 160 kD glycoprotein extracted from desmosomes.[21,23-29] Subsequently, electron microscopic studies with immunogold directly demonstrated that IgG from PF patients bind to the intercellular core of desmosomes in keratinocytes.[30] At this stage it was unclear

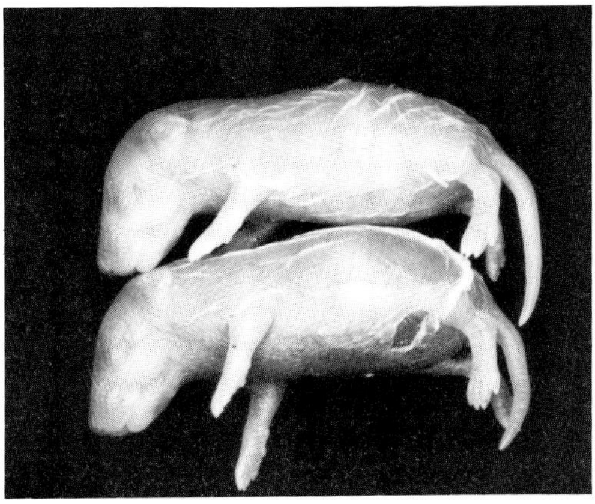

A

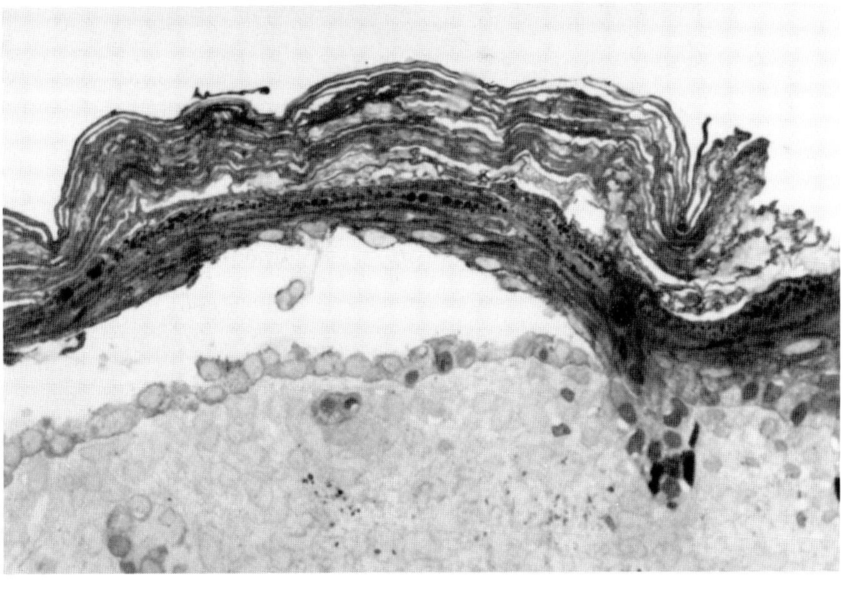

B

Fig. 7.4 Neonatal mice injected with PV IgG demonstrate gross (A) and microscopic (B) blisters similar to lesions seen in patients. (B, x 320).

why PF antibodies only bound stratified squamous epithelia, yet desmoglein was found in the desmosomes of all epithelia and heart.[26,31]

This paradox as well as other biochemical observations about PF antigen were subsequently explained by cloning of desmogleins.[32-35] Basically these cloning studies showed that desmoglein belonged to the cadherin supergene family. This gene family was defined by the cloning of the classical cadherins such as E-, N-, and P-cadherin.[36] These classical cadherins are calcium-dependent cell adhesion molecules whose function has been well established. Desmogleins, like classical cadherins, are transmembrane molecules. Both classical cadherins and desmogleins have homologous amino-terminal extracellular domain structures and share one homologous cytoplasmic subdomain.[37] However desmogleins have a longer carboxy-terminal cytoplasmic tail than do cadherins. This desmoglein cytoplasmic domain contains a desmoglein-specific repeating amino acid motif containing the consensus N-V-V/I-V-T-E-R/S-V-I/V, followed by a carboxy-terminal glycine/serine-rich end subdomain.[37] One striking difference in the cytoplasmic domains is that desmogleins contain cysteines whereas classical cadherins do not.[37] Classical cadherins have six conserved putative calcium binding motifs of acidic amino acids in their extracellular domains.[38-40] Site directed mutagenesis indicates that an amino acid change in one of these sites prevents calcium-binding, changes the conformation, and destroys adhesive function of the molecule.[40] Desmogleins have similar putative calcium binding motifs.[33-35] Finally, cloning has indicated at least three desmoglein genes, DSG1, DSG2, and DSG3 (which encodes PV antigen—see below).[41-43]

Molecular cloning of classical cadherins and desmogleins provides insights into several biochemical observations about PF antigen. The immunochemical finding that most PF antibodies bind calcium-sensitive conformational epitopes is understandable for a molecule in which putative calcium binding sites are thought to stabilize conformation and function. The conserved cysteines in the cytoplasmic domains of desmogleins probably fulfill an important function in binding plakoglobin to PF antigen. Finally, the cloning of DSG1 and DSG2 provide a potential explanation of why PF antibodies only bind stratified squamous epithelia. The DSG1 gene is expressed only in stratified squamous epithelia whereas the DSG2 may be expressed in all desmosome containing tissues.[42,44] In addition, DSG1 is expressed in differentiating layers of stratified squamous epithelia whereas DSG2 is expressed only in the basal layer.[45] PF antigen then would be the gene product of DSG1, accounting for its tissue specific expression.

Characterization of PV antigen yielded a biological surprise, namely that its closest biochemical relative is PF antigen. Immunochemical characterization had indicated that PV antigen is a 130 kD glycoprotein.[20,29,46] Immunoprecipitation studies suggested that PV and PF antigens were related because both bound the 85 kD plakoglobin and both had the same isoelectric point.[22,46] These studies also showed that some PV epitopes were calcium sensitive.[22]

These immunochemical observations were further elucidated when cDNA encoding PV antigen was cloned.[43] Analysis of its amino acid sequence revealed that PV antigen is in the cadherin gene superfamily, most closely related to dsg 1 (the gene product of DSG1, PF antigen). The cytoplasmic domain of PV antigen has a similar subdomain structure as that of dsg 1, except that it lacks the carboxy-terminal glycine/serine-rich end subdomain. It does have both the desmoglein-specific repeating motif (although expressed fewer times than in dsg 1) and, like dsg 1, a subdomain homologous to that seen in the classical cadherins. Finally there are 5 cyteines in conserved locations in the cytoplasmic domains of PV antigen and dsg 1. With this degree of homology in cytoplasmic domains it is not surprising that both PV and PF antigens bind plakoglobin. The extracellular domains of PV and PF antigens are also highly homologous with each other, including the preservation of putative calcium binding domains, explaining the calcium sensitivity of some PV antigen epitopes. Finally there are also four cysteines conserved in similar locations in PV antigen and dsg 1. Cloning, therefore, confirms and extends the close relationship of PV and PF antigens. The PV antigen is now considered a member of the desmoglein subfamily of the cadherin gene superfamily with the gene name DSG3 assigned.[41] Interestingly, all the desmogleins have been localized to human chromosome 18.[47]

If PV antigen is a desmoglein then an obvious question is whether it is also found in desmosomes, as are the other desmogleins. Older immunolocalization studies with patients' skin or antibodies were not conclusive. Some studies showed PV antigen on the entire keratinocyte cell membrane[48-51] whereas others suggested localization to desmosomes.[52-54] The cloning of PV antigen cDNA permitted the preparation of PV antigen-specific antibodies and more definitive immunolocalization studies.[55] Rabbit antibodies raised against fusion proteins containing part of the extracellular domain of PV antigen were used to localize the antigen in cultured keratinocytes. These rabbit antibodies showed intense immunofluorescence staining of the cell surface of these cultured cells (Fig. 7.5B) in a pattern indistinguishable from that of antibodies from PV patients (Fig. 7.5A). This pattern showed dotted fluorescence in tangential membranes, suggestive of junctional (i.e. desmosomal) staining. Ultrastructural local-

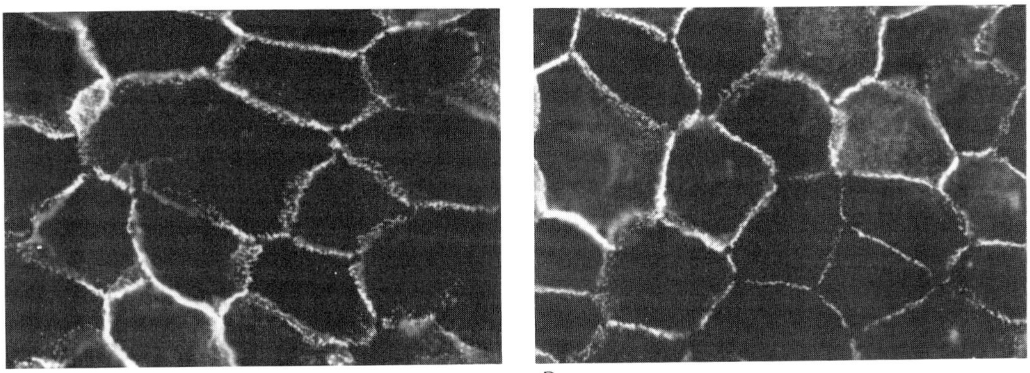

A B

Fig. 7.5 Immunofluorescence of cultured keratinocytes with PV IgG from a patient (A) and antibodies against the extracellular region of PVA raised in rabbits (B). (Magnifications x 630 of original). Reprinted with permission from Kárpáti et al, J Cell Biol 1993; 122:409-415.

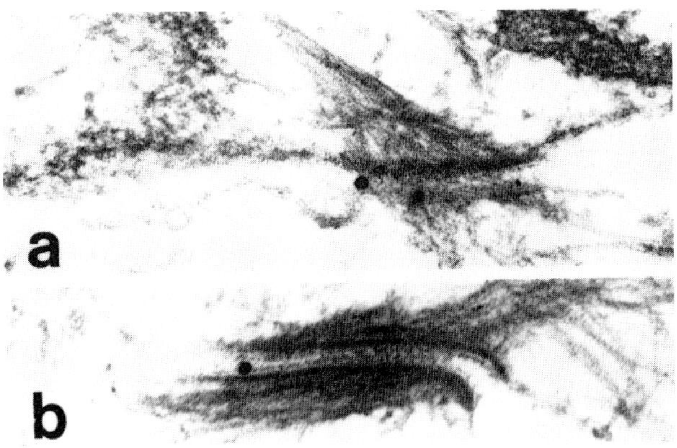

Fig. 7.6 Immunogold electron microscopic localization of PV antigen in cultured human keratinocytes with pre-embedding labeling. PV IgG from patient serum was affinity purified on a fusion protein expressing part of the extracellular domain of PV antigen, and its binding to the cells was identified with silver enhancement of 1 nm gold particles. Note slight binding to closed desmosomes, but no binding along the extradesmosomal cell membrane. (A) and (B) show different desmosomes. (Magnification x 63,000).

ization of this antibody staining by the immunogold technique on paraformaldehyde-fixed cultures (pre-embedding) showed no binding of antibodies to the cell membranes between desmosomes and very slight binding to desmosomes. Figure 7.6 shows such an example with PV IgG affinity purified from a patient's serum on part of the extracellular domain of PV antigen. The findings of intense immunofluorescence staining but minimal immunogold staining suggested that it might be difficult for the gold particles to penetrated intact desmosomes. We, therefore, separated the desmosomes slightly by a brief trypsin/calcium treatment and then fixed and stained the cells. Immunogold staining of such slightly separated desmosomes was strong. Both PV sera and rabbit antibodies stained desmosomes but not interdesmosomal membranes (Fig. 7.7). Similarly, pre-embedding immunogold localization in mouse skin with affinity-purified PV patient sera showed weak but specific desmosomal binding (Fig. 7.8). However, if the affinity-purified antibodies are injected into neonatal mice they cause microscopic separation of desmosomes and there is much more intense immunogold binding to desmosomes, but, again, not to interdesmosomal membranes (Fig. 9). These studies show that, like PF antigen, PV antigen is localized in desmosomes.

The characterization of BP antigen by immunoprecipitation and immunoblotting studies revealed a very surprising and unexpected result—patient sera actually defined two different, unique, and unrelated antigens.[56-62] The antigen now called BPAG1 was estimated by SDS-polyacrylamide gel electrophoresis to be about 220-240 kD. Sera from almost all BP patients recognized this antigen by immunoprecipitation or immunoblotting. Sera from about 50% of BP patients recognize a 180 kD (reported as 166-180 kD) glycoprotein, now called BPAG2, by immunoblotting.[61,63] BPAG2 can also be recognized by immunoprecipitation under certain conditions.[62]

Both BPAG1 and BPAG2 can be identified in preparations of enriched hemidesmosomes.[64,65] Immunoelectron microscopic studies with patient sera showed that their IgG localized to hemidesmosomes,[66-69] both on or near the intracellular plaque and also outside the cell.[70,71] However, since BP sera may bind two different antigens it was not clear if both were

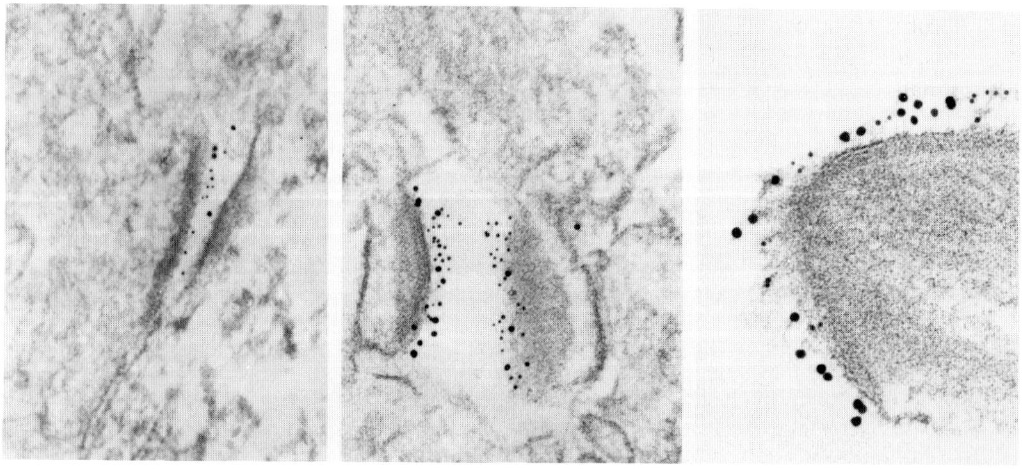

Fig. 7.7 Immunogold electron microscopic localization of PV antigen to separated desmosomes of cultured human keratinocytes that have been briefly treated with trypsin/calcium before fixation. 5 nm gold particles identify rabbit antibodies raised against the extracellular region of PV antigen. 10 nm gold particles identify PV patient IgG. (Left x 75,000; middle x 90,000; right x 150,000 of original). Reprinted with permission from Kárpáti et al, J Cell Biol 1993; 122:409-415.

in the hemidesmosome and if both had epitopes inside and outside the cell. The use of rabbit antibodies, monoclonal antibodies and affinity purified antibodies specific for each of these antigens clarified this confusion. Antibodies against BPAG1 showed it to be inside the cell, associated with the plaque of hemidesmosomes whereas antibodies against BPAG2 showed it to a transmembrane molecule of hemidesmosomes with epitopes inside and outside the cell.[65,72-77]

cDNA cloning further defined these two antigens and proved beyond a doubt that they are biochemically unrelated and from two different genes.[78] Initial studies used human antibodies to screen expression libraries of keratinocyte cDNA.

cDNA encoding part of the carboxy-terminus of BPAG1 was isolated from human and mouse libraries.[69,79,80] Subsequently, the initial cDNA was used for extension cloning to isolate the full length coding sequence for human BPAG1.[81-84] Hybridization studies with the cDNA indicated an mRNA of 9 kb, expressed by keratinocytes but not fibroblasts or other epithelial cells, and encoded by a single human gene.[79,81,85] Southern analysis indicated cross-hybridizing genes in other mammals, but not in birds, amphibia or fish.[85] These findings were confirmed by immunofluorescence detection of BPAG1 in the epidermal basement membrane only of mammals. The human

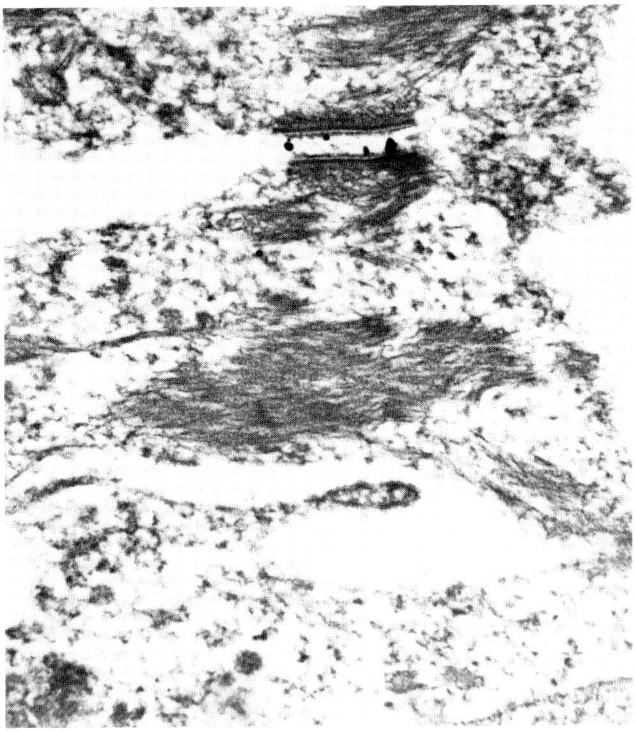

Fig. 7.8 Normal neonatal mouse epidermis stained with PV IgG affinity purified on a fusion protein containing part of the extracellular domain of PV antigen. Pre-embedding labeling with silver enhancement of 1 nm gold shows a few labeled particles in a desmosome, but not elsewhere. (Magnification x 72,000 of original).

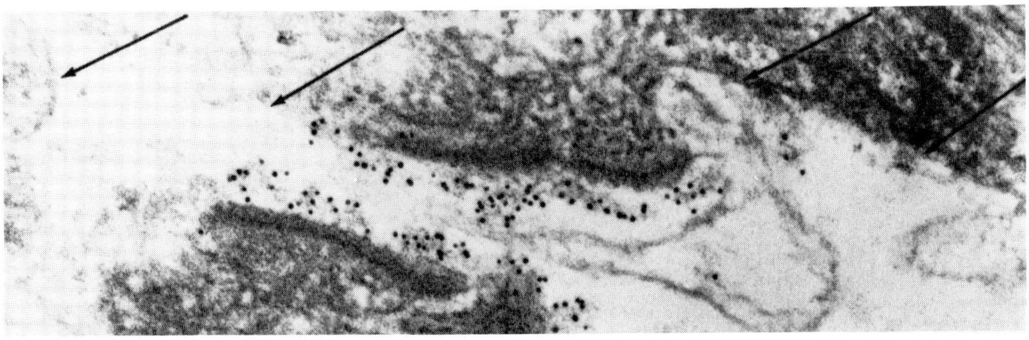

Fig. 7.9 Immunogold localization of PV IgG that was affinity purified from a patient's serum on a fusion protein containing part of the PV antigen extracellular domain and then injected in a neonatal mouse. Note the binding to a separating desmosome but not to the cell membrane between desmosomes (arrows). (Magnification x 127,000 of original). Reprinted with permission from Kárpáti et al, J Cell Biol 1993; 122:409-415.

gene encoding BPAG1 was shown to be on the short arm of chromosome 6.[86-88] Isolation and analysis of the genomic DNA indicated 22 exons in 20 kb of genomic DNA.[86] An upstream tissue-specific promoter/enhancer was also identified.

Cloning of cDNA allowed determination of the amino acid sequence of BPAG1. Database searching for homologous proteins revealed the biologically satisfying result that desmoplakin I is a closely related protein, both in primary sequence and predicted structure (Fig. 7.10).[82,89] This is biologically significant because whereas BPAG1 is in the plaque of the hemidesmosome, desmoplakin I is in the desmosome plaque. Both plaques are thought to fulfill similar functions, namely binding the keratin filaments to these adhesion structures. Subsequently, another intermediate filament binding protein, plectin, was found to be homologous to both these plaque proteins.[90,91]

Analysis of the amino acid sequence of BPAG1 predicted interesting structural and functional features.[82,83] The central part of the molecule displayed a heptad repeat of amino acids with apolar amino acids in the first and fourth position, and a high charge/apolar ratio, features predictive of a two-chain coiled-coil α-helix. The predicted length of this elongated rod domain was greater than 104 nm up to 170 nm.[82,91] Recently, electron microscopy of rotary shadowed BPAG1 has confirmed the presence of a central rod domain of about 140 nm.[92] Desmoplakin I and plectin have this same central coiled-coil rod domain.[93,94] Fast Fourier transform analysis indicates a characteristic repeat of both acidic and basic amino acids with a period of about 10.3-10.5 residues in the rod domain of BPAG1.[82,91] Since the periodicities of the positively and negatively charged residues are out of phase by approximately 180°, they may provide a means of ionic self-aggregation. A further striking homology is that desmoplakin I and plectin have very similar charge periodicities in their rod domains.[90,91]

Additional homologies are found in the carboxy-terminal globular domains of these molecules in which there are homologous subdomains of about 176 amino acids with internal 19 or 38 amino acid repeating units.[91] Desmoplakin I has three of these subdomains, BPAG1 has two, and plectin has six. Fast Fourier transform analysis indicates an acidic and basic charge periodicity in these subdomains of about 9.4,[82,89,90] approximately the same as the charge periodicity found in the 1B rod domain of intermediate filament proteins. This data suggests that BPAG1, desmoplakin I, and plectin may associate with intermediate filaments by their carboxy-terminal regions through charge interactions. Recent studies have confirmed this as-

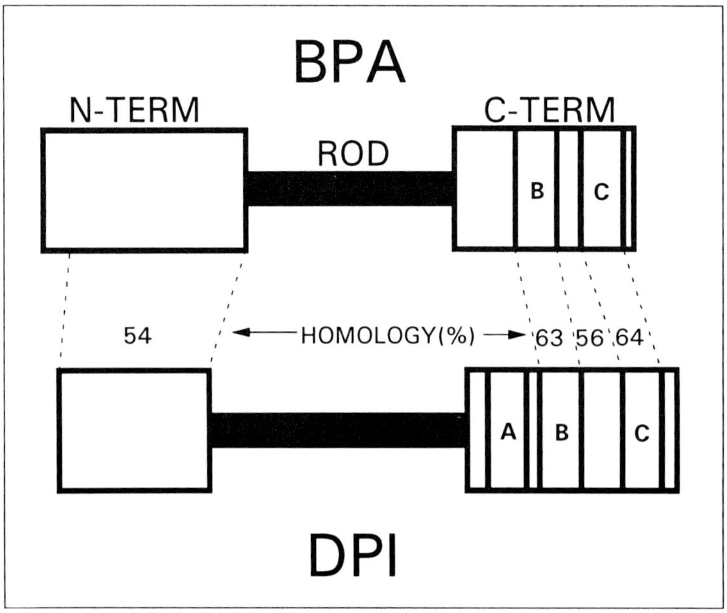

Fig. 7.10 Predicted structure of BP antigen (BPA) and desmoplakin I (DPI) with amino acid sequence homologies. Reprinted with permission from Stanley JR, Adv Immunol 1993; 53:291-325.

sociation for the carboxy-terminal domains of desmoplakin I and plectin.[95,96] In these studies transfection and expression of cDNA encoding these domains shows co-alignment with intermediate filaments. However, whether this co-alignment is due to charge alone or other linking proteins are involved is not known.

Finally, there are also amino acid homologies among the amino-terminal domains of BPAG1, desmoplakin I, and plectin.[91] The amino acid homologies between the amino terminus of BPAG1 and desmoplakin I is shown in Figure 7.11. Notice that the most amino-terminal regions have little, if any, homology, but that the homology starts after about 167 amino acids of desmoplakin and 300 amino acids of BPAG1. This is of biological interest because it has been recently shown that within the first 194 amino acids of desmoplakin I is a desmosomal localization signal.[97] Since most of this area is not homologous in BPAG1 this could provide an explanation of the different localizations of desmoplakin I and BPAG1 to the desmosome and hemidesmosome, respectively.

These homology and structural data, taken together, clearly demonstrate that BPAG1, desmoplakin I, and plectin form a gene family of intermediate filament binding proteins. In this regard it is interesting that both the desmoplakin I and BPAG1 human genes are on chromosome 6p and, therefore, may have resulted from gene duplication.[98]

Both human and mouse BPAG2 cDNAs have also been recently cloned, indicating that BPAG2 clearly does not fall into this same gene family.[76,77,99-102] BPAG2 is a so-called type II transmembrane protein with a collagenous carboxy-terminal extracellular domain and a non-collagenous amino-terminal intracellular domain. The extracellular domain of BPAG2 actually consists of 15 collagen domains of 15 to 242 amino acids each, separated by non-collagenous sequences. BPAG2 is not homologous to other known junctional molecules, but is homologous to a chick corneal collagen. The biological significance of this homology is not clear at this time. The mouse and human BPAG2 amino acid sequences show 87% identity, demonstrating an evolutionarily well-conserved hemidesmosomal component. It has been postulated that the extracellular collagenous domain of BPAG2 may play a key role in adhesion of the basal keratinocyte to the basement membrane (see below).

DO THE AUTOANTIGENS FUNCTION IN CELL ADHESION?

Common wisdom has it that pemphigus and pemphigoid antigens function in cell adhesion. But in science, as in other endeavors, common wisdom may be wrong. The answer is not known at this time, but the evidence on both sides is summarized as follows.

Because patients with pemphigus have autoantibodies that cause loss of cell adhesion it has been assumed that the antigens cause cell adhesion. In fact, pemphigus antigens have been called "intercellular cement substance."[103] Furthermore, as discussed above, pemphigus antigens are members of the cadherin supergene family in which the originally-described members definitely do function in adhesion. In addition, pemphigus antigens are found in desmosomes, which are cell adhesion junctions. Finally, a recent study showed that PV antibodies affinity purified on an area of PVA that is homologous to the homophilic binding domain of E-cadherin caused loss of epidermal cell adhesion when passively transferred to neonatal mice.[104]

Although the above evidence suggests that pemphigus antigens (i.e. desmogleins) might directly mediate cell adhesion there is still no direct evidence to this effect. It has been suggested, for example, that pemphigus antibodies cause loss of cell adhesion by triggering release of proteases by keratinocytes.[105-108] Therefore instead of directly interfering with a presumed adhesive function, autoantibodies might mediate signal transduction across the membrane that leads to protease release that in turn causes the blister.

```
                   .         .         .         .         .
BPAG1     1 MHSSSYSYRSSDSVFSNTTSTRTSLDSNENLLLVHCGPTLINSCISFGSE 50
                                          ::  :   |  ||     :
DP I      1 .............................MYYSRRGVITDQNSDGYCQTG 21

                   .         .         .         .         .
         51 SFDGHRLEMLQQIANRVQRDSVICEDKLILAGNALQSDSKRLESGVQFQN 100
            ::  |      |  |:|     | | |: |  ::| : |
         22 TMSRH......QNQNTIQELLQNCSDCLMRAELIVQPELK.......... 55

                   .         .         .         .         .
        101 EAEIAGYILECENLLRQHVIDVQILIDGKYYQADQLVQRVAKLRDEIMAL 150
                   : :  :| | :       :|  ::||   :  :  |  |:  :
         56 .......YGDGIQLTRSR......ELDECFAQANDQMEILDSLIREMRQM 92

                   .         .         .         .         .
        151 RNECSSVYSKGRILTTEQTKLMISGITQSLNSGFAQTLHPSLTSGLTQSL 200
             : |  :| |  :   || ::::  |:      :
         93 GQPC.DAYQKRLLQLQEQMRALYKAISVPRVRRA................ 125

                   .         .         .         .         .
        201 TPSLTSSSMTSGLSSGMTSRLTPSVTPAYTPGFPSGLVPNFSSGVEPNSL 250
                 ||    |::    |
        126 .....SSKGGGGYTCQSGS............................... 139

                   .         .         .         .         .
        251 QTLKLMQIRKPLLKSSLLDQNLTEEEINMKFVQDLLNWVDEMQVQLDRTE 300
                                    :|:  ::  : | |:    : :|
        140 ......................GWDEFTKHVTSECLGWMRQQRAEMDMVA 167

                   .         .         .         .         .
        301 WGSDLPSVESHLENHKNVHRAIEEFESSLKEAKISEIQMTAPLKLTYAEK 350
            || || ||| |:  |: :|  | ::   |  :|:     :|
        168 WGVDLASVEQHINSHRGIHNSIGDYRWQLDKIKADLREKSA......... 208

                   .         .         .         .         .
        351 LHRLESQYAKLLNTSRNQERHLDTLHNFVSRATNELIWLNEKEEEEVAYD 400
            :  ||  |  ||  |     ||  | |::   : |::|:|: ||||::||
        209 IYQLEEEYENLLKASFERMDHLRQLQNIIQATSREIMWINDCEEEELLYD 258

                   .         .         .         .         .
        401 WSERNTNIARKKDYHAELMRELDQKEENIKSVQEIAEQLLLENHPARLTI 450
            ||::||||| | ::    |  |: ||  :  :    :||:| :|||   |
        259 WSDKNTNIAQKQEAFSIRMSQLEVKEKELNKLKQESDQLVLNQHPASDKI 308

                   .         .         .         .         .
        451 EAYRAAMQTQWSWILQLCQCVEQHIKENTAYFEFFNDAKEATDYLRNLKD 500
            |||   :|||||||||:  |:: |:||| ||| || :|     ||: | |
        309 EAYMDTLQTQWSWILQITKCIDVHLKENAAYFQFFEEAQSTEAYLKGLQD 358

                   .         .         .         .         .
        501 AIQRKYSCDRSSSIHKLEDLVQESMEEKEELLQYKSTIANLMGKAKTIIQ 550
             | :|| ||:   : :| : : |   |:| :| ||  : ||: | | |:|
        359 SIRKKYPCDKNMPLQHLLEQIKELEKEREKILEYKRQVQNLVNKSKKIVQ 408

                   .         .         .         .         .
        551 LKPRNSDCPLKTSIPIKAICDYRQIEITIYKDDECVLANNSHRAKWKVIS 600
            ||||| |      | ::|:|||:|    : | |||:|  |  | || |
        409 LKPRNPDYRSNKPIILRALCDYKQDQKIVHKGDECILKDNNERSKWYVTG 458

                   .         .         .         .         .
        601 PTGNEAMVPSVCFTVPPPNKEAVDLANRIEQQYQNVLTLWHESHINMKSV 650
            | | :::|||| : :||||  ||||  :||| |  :| ||    |||||:
        459 PGGVDMLVPSVGLIIPPPNPLAVDLSCKIEQYYEAILALWNQLYINMKSL 508

                   .         .         .         .         .
        651 VSWHYLINEIDRIRASNVASIKTMLPGEHQQVLSNLQSRFEDFLEDSQES 700
            ||||| : :|::|||  :| :|||   :    :  |  :: :|:  || |
        509 VSWHYCMIDIEKIRAMTIAKLKTMRQEDYMKTIADLELHYQEFIRNSQGS 558
```

Fig. 7.11 Amino acid homologies of amino-termini of BPAG1 (top) and desmoplakin I (DPI) using the GAP program of UWGCG.[115] BPAG1 sequence is from ref. 82-84 and that of DPI is from GenBank accession #M77830, partially referenced by ref. 97.

A recent study tried to determine directly whether the extracellular domain of PVA is capable of substituting for that of E-cadherin in mediating adhesion in transfected L cells (mouse fibroblasts).[109] E-cadherin causes aggregation of these cells that do not normally adhere.[110] However to mediate this adhesion the cytoplasmic domain of E-cadherin must bind α- and β-catenins in these cells, presumably in order to link the transmembrane E-cadherin to the cytoskeleton.[111,112] A chimeric molecule consisting of the extracellular domain of PVA and the transmembrane and cytoplasmic region of E-cadherin was expressed on the cell surface of L cell clones and bound the catenins to the same degree as in E-cadherin transfected cells. However the chimeric molecule only mediated very weak adhesion compared to E-cadherin. These data imply that the extracellular domain of PVA does not mediate strong homophilic adhesion or that other factors are necessary to demonstrate its function.

Similar doubts exist regarding BP antigens' function in adhesion. As discussed above BP antibodies initiate inflammation which may actually lead to the blistering, thus antibodies alone may not be adequate. In addition, BPAG1 is totally inside the cell, therefore it is difficult to comprehend how circulating antibodies could interfere with its function. In contrast, BPAG2 has an extracellular region and an immunodominant epitope, as defined by reactivity of patient sera, has been identified in this domain.[113] In fact, rabbit antibodies raised against the corresponding region in mouse BPAG2 cause epidermal-dermal separation when injected into neonatal mice, suggesting an important pathophysiologic role of this epitope.[19] However, this study does not address whether the antibody interferes directly with a presumed adhesion function or acts through some other mechanism such as recruitment of inflammatory cells whose proteases might damage the adhesive structures at the basement membrane.

THE PRESENT AND THE FUTURE

Clearly the study of autoantibody-mediated blistering diseases has been fruitful in identifying and defining molecules in desmosomes and hemidesmosomes. These diseases also demonstrate, in a dramatic way, the importance of proper functioning of cell adhesion for the maintenance of life. However, an obvious and important question remains unanswered—is the actual cell biologic function of these autoantigens adhesion? This question will undoubtedly occupy cell biologists for some time to come.

REFERENCES

1. Stanley JR. Pemphigus: skin failure mediated by autoantibodies. JAMA 1990; 264:1714-1717.
2. Lever WF. Pemphigus and Pemphigoid. Springfield, IL: Charles C. Thomas, 1965;
3. Beutner EH, Jordon RE, Chorzelski TP. The immunopathology of pemphigus and bullous pemphigoid. J Invest Dermatol 1968; 51:63-80.
4. Beutner EH, Chorzelski TP, Jablonska S. Clinical significance of immunofluorescence tests of sera and skin in bullous diseases: a cooperative study. In: Beutner EH, Chorzelski TP, Kumar V, eds. Immunopathology of the Skin, 3rd ed. New York: John Wiley & Sons, 1987:177-205.
5. Sams WMJr, Jordon RE. Correlation of pemphigoid and pemphigus antibody titres with activity of disease. Br J Dermatol 1971; 84:7-13.
6. Squiquera HL, Diaz LA, Sampaio SAP, et al. Serologic abnormalities in patients with endemic pemphigus foliaceus (fogo selvagem), their relatives, and normal donors from endemic and non-endemic areas of Brazil. J Invest Dermatol 1988; 91:189-191.
7. Merlob P, Metzker A, Hazaz BAC, Rogovin H, Reisner SH. Neonatal pemphigus vulgaris. Pediatrics 1986; 78:1102-1105.
8. Rocha-Alvarez R, Friedman H, Campbell IT, Souza-Aguiar L, Martins-Castro R, Diaz LA. Pregnant women with endemic pem-

phigus foliaceus (Fogo Selvagem) give birth to disease-free babies. J Invest Dermatol 1992; 99:78-82.
9. Schiltz JR, Michel B. Production of epidermal acantholysis in normal human skin in vitro by the IgG fraction from pemphigus serum. J Invest Dermatol 1976; 67:254-260.
10. Hashimoto K, Shafran KM, Webber PS, Lazarus GS, Singer KH. Anti-cell surface pemphigus autoantibody stimulates plasminogen activator activity of human epidermal cells. J Exp Med 1983; 157:259-272.
11. Anhalt GJ, Labib RS, Voorhees JJ, Beals TF, Diaz LA. Induction of pemphigus in neonatal mice by passive transfer of IgG from patients with the disease. N Engl J Med 1982; 306:1189-1196.
12. Roscoe JT, Diaz L, Sampaio SA, et al. Brazilian pemphigus foliaceus autoantibodies are pathogenic to BALB/c mice by passive transfer. J Invest Dermatol 1985; 85:538-541.
13. Anhalt GJ, Till GO, Diaz LA, Labib RS, Patel HP, Eaglstein NF. Defining the role of complement in experimental pemphigus vulgaris in mice. J Immunol 1986; 137:2835-2840.
14. Gammon WR, Lewis DM, Carlo JR, Sams WMJr, Wheeler CEJr. Pemphigoid antibody mediated attachment of peripheral blood leukocytes at the dermal-epidermal junction of human skin. J Invest Dermatol 1980; 75:334-339.
15. Gammon WR, Merritt CC, Lewis DM, Sams WMJr, Wheeler CEJr, Carlo J. Leukocyte chemotaxis to the dermal-epidermal junction of human skin mediated by pemphigoid antibody and complement: mechanism of cell attachment in the in vitro leukocyte attachment method. J Invest Dermatol 1981; 76:514-522.
16. Gammon WR, Merritt CC, Lewis DM, Sams WMJr, Carlo JR, Wheeler CEJr. An in vitro model of immune complex-mediated basement membrane zone separation caused by pemphigoid antibodies, leukocytes, and complement. J Invest Dermatol 1982; 78:285-290.
17. Sams WMJr, Gammon WR. Mechanism of lesion production in pemphigus and pemphigoid. J Am Acad Dermatol 1982; 6:431-452.
18. Stanley JR. Bullous pemphigoid, cicatricial pemphigoid, and chronic bullous disease of childhood. In: Fitzpatrick TB, Eisen AZ, Wolff K, Freedberg IM, Austen KF, eds. Dermatology in General Medicine, 4th ed. New York: McGraw-Hill, 1992.
19. Liu Z, Diaz LA, Troy JL, et al. A passive transfer model of the organ-specific autoimmune disease, bullous pemphigoid, using antibodies generated against the hemidesmosomal antigen, BP180. J Clin Invest 1993; 92:2480-2488.
20. Stanley JR, Koulu L, Thivolet C. Distinction between epidermal antigens binding pemphigus vulgaris and pemphigus foliaceus autoantibodies. J Clin Invest 1984; 74:313-320.
21. Eyre RW, Stanley JR. Human autoantibodies against a desmosomal protein complex with a calcium-sensitive epitope are characteristic of pemphigus foliaceus patients. J Exp Med 1987; 165:1719-1724.
22. Eyre RW, Stanley JR. Identification of pemphigus vulgaris antigen extracted from normal human epidermis and comparison with pemphigus foliaceus antigen. J Clin Invest 1988; 81:807-812.
23. Gorbsky G, Steinberg MS. Isolation of the intercellular glycoproteins of desmosomes. J Cell Biol 1981; 90:243-248.
24. Schmelz M, Duden R, Cowin P, Franke WW. A constitutive transmembrane glycoprotein of Mr 165000 (desmoglein) in epidermal and non-epidermal desmosomes. I. Biochemical identification of the polypeptide. Eur J Cell Biol 1986; 42:177-183.
25. Schmelz M, Duden R, Cowin P, Franke WW. A constitutive transmembrane glycoprotein of Mr 165,000 (desmoglein) in epidermal and non-epidermal desmosomes. II. Immunolocalization and microinjection studies. Eur J Cell Biol 1986; 42:184-199.
26. Cowin P, Garrod DR. Antibodies to epithelial desmosomes show wide tissue and species cross-reactivity. Nature 1983; 302:148-150.
27. Koulu L, Kusumi A, Steinberg MS, Klaus Kovtun V, Stanley JR. Human autoantibodies against a desmosomal core protein in pemphigus foliaceus. J Exp Med 1984;

160:1509-1518.

28. Stanley JR, Koulu L, Klaus Kovtun V, Steinberg MS. A monoclonal antibody to the desmosomal glycoprotein desmoglein I binds the same polypeptide as human autoantibodies in pemphigus foliaceus. J Immunol 1986; 136:1227-1230.
29. Hashimoto T, Ogawa MM, Konohana A, Nishikawa T. Detection of pemphigus vulgaris and pemphigus foliaceus antigens by immunoblot analysis using different antigen sources. J Invest Dermatol 1990; 94:327-331.
30. Rappersberger K, Roos N, Stanley JR. Immunomorphological and biochemical identification of the pemphigus foliaceus autoantigen within desmosomes. J Invest Dermatol 1992; 99:323-330.
31. Rubinstein N, Stanley JR. Pemphigus foliaceus antibodies and a monoclonal antibody to desmoglein I demonstrate stratified squamous epithelial-specific epitopes of desmosomes. Am J Dermatopathol 1987; 9:510-514.
32. Koch PJ, Walsh MJ, Schmelz M, Goldschmidt MD, Zimbelmann R, Franke WW. Identification of desmoglein, a constitutive desmosomal glycoprotein, as a member of the cadherin family of cell adhesion molecules. Eur J Cell Biol 1990; 53:1-12.
33. Goodwin L, Hill JE, Raynor K, Raszi L, Manabe M, Cowin P. Desmoglein shows extensive homology to the cadherin family of cell adhesion molecules. Biochem Biophys Res Commun 1990; 173:1224-1230.
34. Nilles LA, Parry DAD, Powers EE, Angst BD, Wagner RM, Green KJ. Structural analysis and expression of human desmoglein: a cadherin-like component of the desmosome. J Cell Sci 1991; 99:809-821.
35. Wheeler GN, Parker AE, Thomas CL, et al. Desmosomal glycoprotein DGI, a component of intercellular desmosome junctions, is related to the cadherin family of cell adhesion molecules. Proc Natl Acad Sci USA 1991; 88:4796-4800.
36. Takeichi M. Cadherins: a molecular family important in selective cell-cell adhesion. Annu Rev Biochem 1990; 59:237-252.
37. Buxton RS, Magee AI. Structure and interactions of desmosomal and other cadherins. Semin Cell Biol 1992; 3:157-167.
38. Takeichi M. Cadherin cell adhesion receptors as a morphogenetic regulator. Science 1991; 251:1451-1455.
39. Ringwald M, Schuh R, Vestweber D, et al. The structure of cell adhesion molecule uvomorulin. Insights into the molecular mechanism of Ca^{2+}-dependent cell adhesion. EMBO J 1987; 6:3647-3653.
40. Ozawa M, Engel J, Kemler R. Single amino acid substitutions in one Ca^{2+} binding site of uvomorulin abolish the adhesive function. Cell 1990; 63:1033-1038.
41. Buxton RS, Cowin P, Franke WW, et al. Nomenclature of the desmosomal cadherins. J Cell Biol 1993; 121:481-483.
42. Koch PJ, Goldschmidt MD, Zimbelmann R, Troyanovsky R, Franke WW. Complexity and expression patterns of the desmosomal cadherins. Proc Natl Acad Sci U S A 1992; 89:353-357.
43. Amagai M, Klaus-Kovtun V, Stanley JR. Autoantibodies against a novel epithelial cadherin in pemphigus vulgaris, a disease of cell adhesion. Cell 1991; 67:869-877.
44. Schafer S, Koch PJ, Franke WW. Identification of the fundamental human desmoglein, DSG2, and the expression catalogue of a subfamily of desmosomal cadherins. Exp Cell Res 1994; 211: 391-399.
45. Arnemann J, Sullivan KH, Magee AI, King IA, Buxton RS. Stratification-related expression of isoforms of the desmosomal cadherins in human epidermis. J Cell Sci 1993; 104:741-750.
46. Korman NJ, Eyre RW, Klaus-Kovtun V, Stanley JR. Demonstration of an adhering-junction molecule (plakoglobin) in the autoantigens of pemphigus foliaceus and pemphigus vulgaris. N Engl J Med 1989; 321:631-635.
47. Arnemann J, Spurr NK, Buxton RS. The human gene (DSG3) coding for the pemphigus vulgaris antigen is, like the genes coding for the other two known desmogleins, assigned to chromosome 18. Hum Genet 1992; 89:347-350.
48. Wolff K, Schreiner E. Ultrastructural localization of pemphigus auto-antibodies within the epidermis. Nature 1971; 229:59-60.
49. Takigawa M, Imamura S, Ofuji S. Surface

distribution of pemphigus antibody-binding substance(s) on isolated guinea pig epidermal cells. An immunoferritin electron microscopic study. J Invest Dermatol 1978; 71:182-185.

50. Takahashi Y, Patel HP, Labib RS, Diaz LA, Anhalt GJ. Experimentally induced pemphigus vulgaris in neonatal BALB/c mice: a time-course study of clinical, immunologic, ultrastructural, and cytochemical changes. J Invest Dermatol 1985; 84:41-46.
51. Jones JCR, Yokoo KM, Goldman RD. A cell surface desmosome-associated component: identification of a tissue-specific cell adhesion molecule. Proc Natl Acad Sci USA 1986; 83:7282-7286.
52. Jones JCR, Yokoo KM, Goldman RD. Further analysis of pemphigus autoantibodies and their use in studies on the heterogeneity, structure, and function of desmosomes. J Cell Biol 1986; 102:1109-1117.
53. Iwatsuki K, Takigawa M, Imaizumi S, Yamada M. In vivo binding site of pemphigus vulgaris antibodies and their fate during acantholysis. J Am Acad Dermatol 1989; 20:578-582.
54. Akiyama M, Hashimoto T, Sugiura M, Nishikawa T. Ultrastructural localization of pemphigus vulgaris and pemphigus foliaceus antigens in cultured human squamous carcinoma cells. Br J Dermatol 1991; 125:233-237.
55. Karpati S, Amagai M, Prussick R, Cehrs K, Stanley JR. Pemphigus vulgaris antigen, a desmoglein type of cadherin, is localized within keratinocyte desmosomes. J Cell Biol 1993; 122:409-415.
56. Stanley JR, Hawley Nelson P, Yuspa SH, Shevach EM, Katz SI. Characterization of bullous pemphigoid antigen: a unique basement membrane protein of stratified squamous epithelia. Cell 1981; 24:897-903.
57. Stanley JR, Woodley DT, Katz SI. Identification and partial characterization of pemphigoid antigen extracted from normal human skin. J Invest Dermatol 1984; 82:108-111.
58. Labib RS, Anhalt GJ, Patel HP, Mutasim DF, Diaz LA. Molecular heterogeneity of the bullous pemphigoid antigens as detected by immunoblotting. J Immunol 1986; 136:1231-1235.
59. Meyer LJ, Taylor TB, Kadunce DP, Zone JJ. Two groups of bullous pemphigoid antigens are identified by affinity-purified antibodies. J Invest Dermatol 1990; 94:611-616.
60. Cook AL, Hanahoe TH, Mallett RB, Pye RJ. Recognition of two distinct major antigens by bullous pemphigoid sera. Br J Dermatol 1990; 122:435-444.
61. Hashimoto T, Ebihara T, Shimizu H, et al. Comparative study of bullous pemphigoid antigens among Japanese, British, and U.S. patients indicates similar antigen profiles with the 170-kD antigen present both in the basement membrane and on the keratinocyte cell membrane. J Invest Dermatol 1993; 100:385-389.
62. Ebihara T, Hashimoto T, Kudoh J, Gamou S, Shimizu N, Nishikawa T. Detection of the 170-kDa bullous pemphigoid antigen by immunoprecipitation. J Invest Dermatol 1993; 100:176-179.
63. Morrison LH, Labib RS, Zone JJ, Diaz LA, Anhalt GJ. Herpes gestationis autoantibodies recognize a 180-kD human epidermal antigen. J Clin Invest 1988; 81:2023-2026.
64. Owaribe K, Nishizawa Y, Franke WW. Isolation and characterization of hemidesmosomes from bovine corneal epithelial cells. Exp Cell Res 1991; 192:622-630.
65. Nishizawa Y, Uematsu J, Owaribe K. HD4, a 180 kDa bullous pemphigoid antigen, is a major transmembrane glycoprotein of the hemidesmosome. J Biochem (Tokyo) 1993; 113:493-501.
66. Mutasim DF, Takahashi Y, Labib RS, Anhalt GJ, Patel HP, Diaz LA. A pool of bullous pemphigoid antigen(s) is intracellular and associated with the basal cell cytoskeleton-hemidesmosome complex. J Invest Dermatol 1985; 84:47-53.
67. Westgate GE, Weaver AC, Couchman JR. Bullous pemphigoid antigen localization suggests an intracellular association with hemidesmosomes. J Invest Dermatol 1985; 84:218-224.
68. Regnier M, Vaigot P, Michel S, Prunieras M. Localization of bullous pemphigoid antigen (BPA) in isolated human keratinocytes. J Invest Dermatol 1985; 85:187-190.

69. Owaribe K, Kartenbeck J, Stumpp S, et al. The hemidesmosomal plaque. I. Characterization of a major constituent protein as a differentiation marker for certain forms of epithelia. Differentiation 1990; 45:207-220.
70. Mutasim DF, Morrison LH, Takahashi Y, et al. Definition of bullous pemphigoid antibody binding to intracellular and extracellular antigen associated with hemidesmosomes. J Invest Dermatol 1989; 92:225-230.
71. Shimizu H, McDonald JN, Kennedy AR, Eady RAJ. Demonstration of intra- and extracellular localization of bullous pemphigoid antigen using cryofixation and freeze substitution for postembedding immunoelectron microscopy. Arch Dermatol Res 1989; 281:443-448.
72. Tanaka T, Korman NJ, Shimizu H, et al. Production of rabbit antibodies against carboxy-terminal epitopes encoded by bullous pemphigoid cDNA. J Invest Dermatol 1990; 94:617-623.
73. Shimizu H, Hashimoto T, Nishikawa T, Eady RAJ. Human monoclonal anti-basement membrane zone antibodies derived from virally transformed lymphocytes of a patient with bullous pemphigoid recognize epitopes associated with hemidesmosomes. Br J Dermatol 1991; 124:217-220.
74. Ishiko A, Shimizu H, Kikuchi A, Ebihara T, Hashimoto T, Nishikawa T. Human autoantibodies against the 230-kD bullous pemphigoid antigen (BPAG1) bind only to the intracellular domain of the hemidesmosome, whereas those against the 180-kD bullous pemphigoid antigen (BPAG2) bind along the plasma membrane of the hemidesmosome in normal human and swine skin. J Clin Invest 1993; 91:1608-1615.
75. Klatte DH, Kurpakus MA, Grelling KA, Jones JCR. Immunochemical characterization of three components of the hemidesmosome and their expression in cultured epithelial cells. J Cell Biol 1989; 109:3377-3390.
76. Diaz L, Ratrie H, Saunders WS, et al. Isolation of a human epidermal cDNA corresponding to the 180-kD autoantigen recognized by bullous pemphigoid and herpes gestationis sera. J Clin Invest 1990; 86:1088-1094.
77. Hopkinson SB, Riddelle KS, Jones JC. Cytoplasmic domain of the 180-kD bullous pemphigoid antigen, a hemidesmosomal component: molecular and cell biologic characterization. J Invest Dermatol 1992; 99:264-270.
78. Sawamura D, Li K, Uitto J. 230-kD and 180-kD bullous pemphigoid antigens are distinct gene products (letter). J Invest Dermatol 1992; 98:942-943.
79. Stanley JR, Tanaka T, Mueller S, Klaus-Kovtun V, Roop D. Isolation of cDNA for bullous pemphigoid antigen by use of patients' autoantibodies. J Clin Invest 1988; 82:1864-1870.
80. Amagai M, Hashimoto T, Tajima S, et al. Partial cDNA cloning of the 230-kD mouse bullous pemphigoid antigen by use of a human monoclonal anti-basement membrane zone antibody. J Invest Dermatol 1990; 95:252-259.
81. Sawamura D, Li KH, Nomura K, Sugita Y, Christiano AM, Uitto J. Bullous pemphigoid antigen: cDNA cloning, cellular expression, and evidence for polymorphism of the human gene. J Invest Dermatol 1991; 96:908-915.
82. Tanaka T, Parry DAD, Klaus-Kovtun V, Steinert PM, Stanley JR. Comparison of molecularly cloned bullous pemphigoid antigen to desmoplakin I confirms that they define a new family of cell adhesion junction plaque proteins. J Biol Chem 1991; 266:12555-12559.
83. Sawamura D, Li K, Chu ML, Uitto J. Human bullous pemphigoid antigen (BPAG1). Amino acid sequences deduced from cloned cDNAs predict biologically important peptide segments and protein domains. J Biol Chem 1991; 266:17784-17790.
84. Elgart GW, Stanley JR. Cloning of the 5' mRNA for the 230-kD bullous pemphigoid antigen by rapid amplification of cDNA ends. J Invest Dermatol 1993; 101:244-246.
85. Amagai M, Elgart GW, Klaus-Kovtun V, Stanley JR. Southern analysis of the 230-kD bullous pemphigoid antigen gene in normal humans, animals, and patients with junctional epidermolysis bullosa. J Invest

Dermatol 1991; 97:249-253.
86. Tamai K, Sawamura D, Do HC, Tamai Y, Li K, Uitto J. The human 230-kDa bullous pemphigoid antigen (BPAG1): exon-intron organization and identification of regulatory tissue specific elements in the promoter region. J Clin Invest 1993; 92:814-822.
87. Sawamura D, Nomura K, Sugita Y, et al. Bullous pemphigoid antigen (BPAG1): cDNA cloning and mapping of the gene to the short arm of human chromosome 6. Genomics 1990; 8:722-726.
88. Minoshima S, Amagai M, Kudoh J, et al. Localization of the human gene for 230-kDal bullous pemphigoid autoantigen (BPAG1) to chromosome 6pter—q15. Cytogenet Cell Genet 1991; 57:30-32.
89. Green KJ, Parry DAD, Steinert PS, et al. Structure of the human desmoplakins. Implications for function in the desmosomal plaque. J Biol Chem 1990; 265:2603-2612.
90. Wiche G, Becker B, Luber K, et al. Cloning and sequencing of rat plectin indicates a 466-kD polypeptide chain with a three-domain structure based on a central alpha-helical coiled coil. J Cell Biol 1991; 114:83-99.
91. Green KJ, Virata MLA, Elgart GW, Stanley JR, Parry DAD. Comparative structural analysis of desmoplakin, bullous pemphigoid antigen and plectin: members of a new gene family involved in organization of intermediate filments. Int J Biol Macrobiol 1992; 14:145-153.
92. Klatte DH, Jones JCR. Purification of the 230-kD bullous pemphigoid antigen (BP230) from bovine tongue mucosa: structural analyses and assessment of BP230 tissue distribution using a new mono-clonal antibody. J Invest Dermatol 1994; 102:39-44.
93. O'Keefe EJ, Erickson HP, Bennett V. Desmoplakin I and desmoplakin II. Purification and characterization. J Biol Chem 1989; 264:8310-8318.
94. Foisner R, Feldman B, Sander L, Wiche G. Monoclonal antibody mapping of structural and functional plectin epitopes. J Cell Biol 1991; 112:397-405.
95. Stappenbeck TS, Green KJ. The desmoplakin carboxyl terminus coaligns with and specifically disrupts intermediate filament networks when expressed in cultured cells. J Cell Biol 1992; 116:1197-1209.
96. Wiche G, Gromov D, Donovan A, Castanon MJ, Fuchs E. Expression of plectin mutant cDNA in cultured cells indicates a role of COOH-terminal domain in intermediate filament association. J Cell Biol 1993; 121:607-619.
97. Stappenbeck TS, Bornslaeger EA, Corcoran CM, Luu HH, Virata ML, Green KJ. Functional analysis of desmoplakin domains: specification of the interaction with keratin versus vimentin intermediate filament networks. J Cell Biol 1993; 123:691-705.
98. Arnemann J, Spurr NK, Wheeler GN, Parker AE, Buxton RS. Chromosomal assignment of the human genes coding for the major proteins of the desmosome junction, desmoglein DGI (DSG), desmocollins DGII/III (DSC), desmoplakins DPI/II (DSP), and plakoglobin DPIII (JUP). Genomics 1991; 10:640-645.
99. Giudice GJ, Squiquera HL, Elias PM, Diaz LA. Identification of two collagen domains within the bullous pemphigoid autoantigen, BP180. J Clin Invest 1991; 87:734-738.
100. Giudice GJ, Emery DJ, Diaz LA. Cloning and primary structural analysis of the bullous pemphigoid autoantigen, BP180. J Invest Dermatol 1992; 99:243-250.
101. Li K, Guidice GJ, Tamai K, et al. Cloning of partial cDNA for mouse 180-kDa bullous pemphigoid antigen (BPAG2), a highly conserved collagenous protein of the cutaneous basement membrane zone. J Invest Dermatol 1992; 99:258-263.
102. Li K, Tamai K, Tan EM, Uitto J. Cloning of type XVII collagen. Complementary and genomic DNA sequences of mouse 180-kilodalton bullous pemphigoid antigen (BPAG2) predict an interrupted collagenous domain, a transmembrane segment, and unusual features in the 5'-end of the gene and the 3'-untranslated region of the mRNA. J Biol Chem 1993; 268:8825-8834.
103. Hashimoto K, Lever WF. An electron microscopic study on pemphigus vulgaris of the mouth and the skin with special reference to the intercellular cement. J Invest Dermatol 1967; 48:540-552.

104. Amagai M, Karpati S, Prussick R, Klaus-Kovtun V, Stanley JR. Autoantibodies against the amino-terminal cadherin-like binding domain of pemphigus vulgaris antigen are pathogenic. J Clin Invest 1992; 90:919-926.
105. Farb RM, Dykes R, Lazarus GS. Anti-epidermal-cell-surface pemphigus antibody detaches viable epidermal cells from culture plates by activation of proteinase. Proc Natl Acad Sci USA 1978; 75:459-463.
106. Schiltz JR, Michel B, Papay R. Pemphigus antibody interaction with human epidermal cells in culture. J Clin Invest 1978; 62:778-788.
107. Schiltz JR, Michel B, Papay R. Appearance of "pemphigus acantholysis factor" in human skin cultured with pemphigus antibody. J Invest Dermatol 1979; 73:575-581.
108. Morioka S, Lazarus GS, Jensen PJ. Involvement of urokinase-type plasminogen activator in acantholysis induced by pemphigus IgG. J Invest Dermatol 1987; 89:474-477.
109. Amagai M, Karpati S, Klaus-Kovtun V, Udey MC, Stanley JR. The extracellular domain of pemphigus vulgaris antigen (desmoglein 3) mediates weak homophilic adhesion. J Invest Dermatol 1994; 102:402-408.
110. Nagafuchi A, Shirayoshi Y, Okazaki K, Yasuda K, Takeichi M. Transformation of cell adhesion properties by exogenously introduced E-cadherin cDNA. Nature 1987; 329:341-343.
111. Nagafuchi A, Takeichi M. Cell binding function of E-cadherin is regulated by the cytoplasmic domain. EMBO J 1988; 7:3679-3684.
112. Ozawa M, Ringwald M, Kemler R. Uvomorulin-catenin complex formation is regulated by a specific domain in the cytoplasmic region of the cell adhesion molecule. Proc Natl Acad Sci U S A 1990; 87:4246-4250.
113. Giudice GJ, Emery DJ, Zelickson BD, Anhalt GJ, Liu Z, Diaz LA. Bullous pemphigoid and herpes gestationis autoantibodies recognize a common non-collagenous site on the BP180 ectodomain. J Immunol 1993; 151:5742-5750.
114. Stanley JR. Cell adhesion molecules as targets of autoantibodies in pemphigus and pemphigoid, bullous diseases due to defective epidermal cell adhesion. Adv Immunol 1993; 53:291-325.
115. Devereux J, Haeberli P, Smithies O. A comprehensive set of sequence analysis programs for the VAX. Nucleic Acids Res 1984; 12:387-395.

INDEX

M

N

O

P

R

S

T